A FUNCTIONAL BIOLOGY OF
FREE-LIVING PROTOZOA

A Functional Biology of Free-Living Protozoa

JOHANNA LAYBOURN-PARRY, BSc, MSc, PhD

*Department of Biological Sciences,
University of Lancaster, Lancaster, UK*

UNIVERSITY OF CALIFORNIA PRESS
Berkeley and Los Angeles

©1984 Johanna Laybourn-Parry

University of California Press
Berkeley and Los Angeles, California

Library of Congress Cataloging in Publication Data

Laybourn-Parry, Johanna.
  A functional biology of free-living protozoa.

  (Functional biology series)
  Bibliography: p
  Includes index.
  1. Protozoa. I. Title. II. Series.
  QL366.L25   1984   593.1   84-2569
  ISBN 0-520-05339-7
  ISBN 0-520-05340-0 (pbk.)

56,967

CONTENTS

CAMROSE LUTHERAN COLLEGE v
LIBRARY

FUNCTIONAL BIOLOGY SERIES: FOREWORD

General Editor: Peter Calow, Department of Zoology,
University of Sheffield, England

The main aim of this series will be to illustrate and to explain the way organisms 'make a living' in nature. At the heart of this — their *functional biology* — is the way organisms acquire and then make use of resources in metabolism, movement, growth, reproduction, and so on. These processes will form the fundamental framework of all the books in the series. Each book will concentrate on a particular taxon (species, family, class or even phylum) and will bring together information on the form, physiology, ecology and evolutionary biology of the group. The aim will be not only to describe *how* organisms work, but also to consider *why* they have come to work in that way. By concentrating on taxa which are well known, it is hoped that the series will not only illustrate the success of selection, but also show the constraints imposed upon it by the physiological, morphological and developmental limitations of the groups.

Another important feature of the series will be its *organismic orientation*. Each book will emphasise the importance of functional *integration* in the day-to-day lives and the evolution of organisms. This is crucial since, though it may be true that organisms can be considered as collections of gene-determined traits, they nevertheless interact with their environment as integrated wholes and it is in this context that individual traits have been subjected to natural selection and have evolved.

The key features of the series are, therefore:

(1) Its emphasis on whole organisms as integrated, resource-using systems.
(2) Its interest in the way selection and constraints have moulded the evolution of adaptations in particular taxonomic groups.
(3) Its bringing together of physiological, morphological, ecological and evolutionary information.

This volume, on free-living Protozoa, is the first in the series.

P. Calow

PREFACE AND ACKNOWLEDGEMENTS

One feature of protozoological research which struck me from the beginning of my association with the Protozoa as a researcher, was that the study of these organisms is often divorced from the natural environment. A considerable number of species lend themselves to easy culture in the laboratory and some have become familiar laboratory organisms. There is a wealth of information on such species, which represent only an infinitesimal portion of the representatives of the protozoan sub-kingdom which inhabit the waters and soils of the world. Inevitably organisms routinely maintained in culture tend to be studied in the context of the laboratory, and although the data generated are interesting and often extremely valuable, they frequently lack an essential component: they tell us very little about the organism as it functions in nature. We can never hope to gain a full understanding of an organism's functional biology unless we consider how the chemical and physical characteristics of its natural environment can modify the performance of essential life processes. There are relatively few free-living organisms which enjoy a constant environment, most face short- or long-term changes in temperature, pH, food supply, oxygen availability, moisture and light. The impact of such variations on the physiology of an organism will to a large extent be a function of its physiological and ecological tolerances, and its adaptability. Thus the ideal approach to the study of the Protozoa should take account of the variable conditions prevailing in the natural environment. Moreover, this approach may ultimately give us an insight into how various patterns of biological function have evolved in the Protozoa.

Various people have given me help in the preparation of this book and it is with pleasure that I take this opportunity to thank them. My thanks are due to Christine Martin and Margaret Holden who typed the manuscript, Christine Kingsmill who helped with the bibliography and Ken Oates for his invaluable help with photography. I am particularly grateful to those publishers and authors, acknowledged in detail elsewhere, who allowed me to reproduce graphs and illustrations. Many were kind enough to send me original drawings and photographs. My thanks must also go to those who offered valuable criticism in the preparation of this book and the general editor of the Functional Biology series, Dr Peter Calow, for his guidance.

1 THE PROTOZOAN CELL

A. Introduction

The unicellular organisms which constitute the Protozoa exhibit considerable morphological and physiological diversity. Although the majority resemble members of the Animal Kingdom in their feeding behaviour, some members of the Protozoa possess the fundamental plant characteristic of autotrophic nutrition. In a considerable number of species possessing chloroplasts and an autotrophic capacity, some degree of heterotrophy must also be practised, so that some protozoans sit on the boundary between a plant-like nutrition and an animal-like nutrition.

Protozoa are eukaryote organisms and have long been referred to as unicellular, but their complexity has inevitably led to debate as to their unicellular status. Baker (1948) pointed out that since many Protozoa display the characteristics of a cell so well, it is difficult to understand how their unicellular status can be denied. However he argues that since some members of the Protozoa, notably the ciliates, are polyploid, they cannot be compared to the single haploid or diploid cells of higher organisms. In a more recent consideration of the question Corliss (1972) makes the point that the ultrastructural similarities between protozoans and the cells of multicellular organisms make a clear case for considering the Protozoa as single-celled organisms. Protozoan cells are more complex than the cells of metazoans, as one would expect, when one considers the long presence of the Protozoa in evolutionary history and the wide range of functions which must be performed by a unicellular entity.

Although the larger species can be seen with the naked eye, no protozoan cell can be looked at in any detail without the aid of a microscope; thus by definition the Protozoa are microorganisms. The microscope gives the protozoologist his entry into the strange, complex and often startlingly beautiful microscopic world, inhabited by an array of interacting organisms including the Protozoa. The first man to survey and describe this exciting microscopic domain, was a Dutch draper called Antony van Leeuwenhoek (1632-1723). As a consequence he is often called the father of protozoology. His detailed observations of what he called his animalicules, were made with simple lenses which he ground himself, and with which he was able to achieve magnifications of 200-300 times. The painstaking and elegant observations of van Leeuwenhoek were reported in Dutch to the Royal Society of London. Since that time the study of protozoology has progressed to the sophisticated physiological, ultrastructural and biochemical science it is today.

The ease with which some protozoan species can be cultured, and the fact that they are single cells, has resulted in Protozoa becoming the subject of experiments, which although intrinsically interesting, do nothing to elucidate the functional biology of these creatures in their natural environment. The aim of this book is to consider the biology of protozoan cells as they perform in their natural environment, rather than the aberrant world of the laboratory incubator. The chemical and physical characteristics of the natural environment impose constraints on the physiology of organisms; and all organisms interact with other biotic components of the community in which they live. The impact of both biotic and abiotic factors on the physiological performance and the population dynamics of a species are important to our understanding of the role and position of any population in nature. Much of the understanding of the physiological and biochemical functioning of Protozoa has of necessity been derived from laboratory experiments. The ideal experimental approach attempts to reflect, as far as possible, the environmental and biological conditions of the natural habitat in the laboratory, so that the information generated has an applicability to our understanding of how wild protozoan populations and communities function.

B. Protozoan Systematics

The Protozoa are not a natural group: they are simply a collection of single-celled eukaryotes placed together for convenience. In the traditional classification of living organisms the Protozoa were placed as a phylum in the Animal Kingdom. The autotrophic species carrying photosynthetic pigments, however, were also classified among the algae by botanists. In many ways the old classification of the living world into Animal and Plant Kingdoms was unsatisfactory because of such anomalies. The new five-kingdom classification (Whittaker, 1969; Margulis, 1974) is now widely accepted. In this new system the living world is divided into Monera, Protista, Plantae, Fungi and Animalia. The Protozoa are divided into a number of phyla along with other organisms in the kingdom Protista. Margulis (1974) has described the protists as a miscellaneous group of phyla in which profound 'evolutionary experimentation' has occurred. Their status as a natural group is accepted because during their diversification mitosis and meiosis became established. Some members of this diverse group eventually became hosts to photosynthetic prokaryotes which in time evolved into plastids (Margulis, 1974).

The five-kingdom system of classification is consistent with the cell symbiosis theory, which contends that during the early and mid-Precambrian prokaryote organisms originated and diversified and gave rise to eukaryotes by a series of specific symbioses during the later Precambrian. The process of hereditary endosymbiosis in which one type of cell becomes an intracellular self-reproducing inhabitant of another type of cell, is regarded as the major evolutionary mechanism in the development of certain eukaryotic organelles, such as chloroplasts and mitochondria (Margulis, 1970, 1981).

The fact that the artificial group called the Protozoa are now reclassified into the Protista has meant that many groups previously holding the status of classes in the old classification are now elevated to the status of phyla. The most recent complete revision of the Protozoa, the product of the Committee of Systematics and Evolution of the Society of Protozoologists (Levine et al., 1980) appeared in 1980 and superseded the 1964 classification system (Honigberg et al., 1964). The new classification is shown in an abbreviated form in Table 1.1 and it should be noted that it is one of convenience and does not necessarily indicate evolutionary relationships. Moreover, since it is the product of a committee, there are bound to be areas of disagreement. Essentially the new system presents something which protozoologists can use and modify as necessary with new developments in the field of protozoan systematics. Indeed it is worth noting that protozoan systematics are in a fluid state, each year new findings being published which necessitate a revision in the position of various orders and genera. The fact that the Society of Protozoology has a committee specifically to consider systematics highlights this very problem.

Table 1.1: Revised Classification of the Sub-kingdom Protozoa, after Levine *et al.* (1980)

KINGDOM   PROTISTA   SUB-KINGDOM   PROTOZOA

PHYLUM 1	SARCOMASTIGOPHORA (one type of nucleus, except in heterokaryotic Foraminiferida; flagella or pseudopodia or both)
Subphylum I	MASTIGOPHORA (one or more flagella; one type of nucleus; binary fission; autotrophic or heterotrophic)
Class I	PHYTOMASTIGOPHOREA (usually with chloroplasts; those lacking chloroplasts show a clear relationship with pigmented forms; most free-living)
Orders	Cryptomonadida Dinoflagellida Euglenida Chrysomonadida Heterochlorida Chloromonadida Prymesiida Volvocida Prasinomonadida Silicoflagellida
Class 2	ZOOMASTIGOPHOREA (chloroplasts absent; one to many flagella; some amoeboid forms; sexuality known in some groups; polyphyletic group)
Orders	Choanoflagellida Kinetoplastida (mostly parasitic) Proteromonadida (parasitic) Retortamonadida (parasitic) Diplomonadida (free-living and parasitic) Oxymonadida (parasitic) Trichomonadida (mostly parasitic) Hypermastigida (parasitic)
Subphylum II	OPALINATA (numerous cilia; no cytostome; binary fission; syngamy with anisogamous flagellated gametes; all parasitic)
Subphylum III	SARCODINA (pseudopodia, or locomotion without discrete pseudopodia achieved by protoplasmic flow; naked or with internal or external skeleton;

Table 1.1: Contd.

	binary fission; sexuality not common but if present involving flagellate, or rarely amoeboid, gametes; mainly free-living)
Superclass 1	RHIZOPODA (locomotion by lobopodia, filopodia or reticulopodia, or protoplasmic flow without the formation of discrete pseudopodia)
Class 1	LOBOSEA
Subclass 1	GYMNAMOEBIA
Orders	Amoebida
	Schizopyrenida
	Pelobiontida
Subclass 2	TESTACEALOBOSIA (body enclosed in a test; tectum or other membrane external to the plasma membrane)
Orders	Arcellinida
	Trichosida
Class 2	ACARPOMYXEA (small plasmodia or much expanded similar uninucleate forms, usually branching, sometimes forming reticulum of branches; no test, no spores)
Orders	Heptomyxida
	Stereomyxida
Class 3	ACRASEA (uninucleate; eruptive lobose pseudopodia; sexuality unknown)
Orders	Acrasida
Class 4	EUMYCETOZOEA (Myxamoebae with filiform subpseudopodia, flagella sometimes present; producing aerial fruiting bodies)
Subclass 1	PROTOSTELIIA
Orders	Protosteliida
Subclass 2	DICTYOSTELIIA
Orders	Dictyosteliida
Subclass 3	MYXOGASTRIA
Orders	Echinosteliida
	Liceida
	Trichiida
	Stemonitida
	Physarida
Class 5	PLASMODIOPHOREA (all parasitic)
Orders	Plasmodiophorida

Table 1.1: Contd.

Class 6	FILOSEA (hyaline filiform pseudopodia, often branching)
Orders	Aconchulinida Gromiida
Class 7	GRANULORETICULOSEA (usually delicate finely granular or hyaline reticulopodia)
Orders	Athalamida Monothalamida Foraminiferida
Class 8	XENOPHYOPHOREA (multinucleate plasmodium enclosed in a branched-tube system made up of organic material)
Orders	Psamminida Stannomida
Superclass 2	ACTINOPODA (often spherical; axopodia with microtubular stereoplasm; skeleton composed of organic matter and/or silica or strontium sulphate sometimes present; asexual and/or sexual reproduction)
Class 1	ACANTHAREA (strontium sulphate skeleton composed of variously arranged spines)
Orders	Holocanthida Symphyacanthida Chaunacanthida Arthracanthida Actineliida
Class 2	POLYCYSTINEA (siliceous skeleton usually consisting of latticed shells with or without radial spines, axonemes arising from axoplast in endoplasm)
Orders	Spumellarida Nassellarida
Class 3	PHAEODAREA (skeleton — sometimes absent — of mixed silica and organic matter)
Orders	Phaeocystida Phaeospharerida Phaeocalpida Phaeogromida Phaeoconchida Phaeodendrida

Table 1.1: Contd.

Class 4	HELIOZOEA (without central capsule, skeletal structures of silica or organic material if present; axopodia radiating on all sides)
Orders	Desmothoracida Actinophryida Taxopodida Centrohelida
PHYLUM II	LABYRINTHOMORPHA (saprobic and parasitic on algae)
PHYLUM III	APICOMPLEXA (all parasitic)
PHYLUM IV	MICROSPORA (all parasitic)
PHYLUM V	ASCETOSPORA (all parasitic)
PHYLUM VI	MYXOZOA (all parasitic)
PHYLUM VII	CILIOPHORA (two types of nuclei; simple cilia or compound ciliary organelles; binary fission, multiple fission or budding; sexual reproduction by conjugation, autogamy and cytogamy; mostly free-living)
Class 1	KINETOFRAGMINOPHOREA (oral infraciliature only slightly distinct from somatic infraciliature; cytostome apical or sub- apical on body surface or at bottom of atrium or vestibulum)
Subclass 1 Orders	GYMNOSTOMATIA Prostomatida Pleurostomatida Primociliatida Karyorelictida (There is some controversy over the inclusion of the latter two orders in the subclass GYMNOSTOMATIA — ultrastructural studies indicate flagellate characteristics from some members.)
Subclass 2	VESTIBULIFERIA (free-living or parasitic; apical or rear apical

Table 1.1: Contd.

	vestibulum commonly present, equipped with cilia derived from anterior somatic kinetics leading to cytostome)
Orders	Trichostomatida
	Entodiniomorphida
	Colpodida
Subclass 3	HYPOSTOMATIA
Superorder 1	Nassulidea
Orders	Synhymeniida
	Nassulida
Superorder 2	Phyllopharyngidea
Orders	Cyrtophorida
	Chonotrichida
Superorder 3	Rhynchodea
Orders	Rhynchodida
Superorder 4	Apostomatidea
Orders	Apostomatida
Subclass 4	SUCTORIA
Orders	Suctorida
Class 2	OLIGOHYMENOPHOREA (oral apparatus at least partially in buccal cavity; oral ciliature distinct from somatic ciliature; some species lorcate; colony formation common in some groups)
Subclass 1	HYMENOSTOMATIA (body ciliature often uniform and heavy)
Orders	Hymenostomatida
	Scuticociliatida
	Astomatida
Subclass 2	PERITRICHIA
Order 1	Peritrichida
Class 3	POLYMENOPHOREA (well-developed, conspicuous adoral zone; somatic ciliature complete or reduced appearing as cirri; cytostome in buccal cavity)
Subclass 1	SPIROTRICHA (characters of class)
Orders	Heterotrichida
	Odontostomatida
	Oligotrichida
	Hypotrichida

C. Evolution of Protozoa

It has been pointed out by Corliss (1961) with regard to ciliates, but is true of the majority of Protozoa, that there are immense difficulties in tracing evolutionary affinities because obstacles exist, which do not exist in metazoan studies. Most Protozoa do not produce fossil remains, the only exceptions being some Sarcodina which possess mineral skeletal structures. The microscopic size of protozoans and their sub-cellular organisation makes study difficult. In addition they are cosmopolitan and there is frequently a lack of recognisable sexuality. All these factors compound the problem of determining the affinities of the various groups within the sub-kingdom. Corliss (1961) argues that since we can assume with certainty that the present ciliate groups have survived from early geological time relatively unchanged, the present Protozoa must closely resemble the groups from which they have arisen. Thus careful comparative study of extant forms can overcome the problem of a poor fossil record without eliminating the dimension of time.

In the past there has been dispute as to whether the Protozoa are monophyletic or polyphyletic in origin. Grassé (1952) believed them to be monophyletic and derived from bacteria. However, it is now widely accepted that the Protozoa are a diverse group of polyphyletic origin, the name Protozoa indicating a level of organisation rather than an evolutionary relationship (Kerkut, 1960; Whittaker, 1977).

Deciding which of the Protozoa are the most primitive is difficult and has proved a matter of some controversy. During the early part of this century protozoologists had only morphological criteria on which to base systematic hypotheses. Not unreasonably it was assumed that the Sarcodina contained the most primitive Protozoa, since a shapeless piece of protoplasm is less morphologically advanced than a fixed organised cell shape, which may possess locomotory organelles such as flagella or cilia. With the development of the modern physiological and biochemical approach to protozoology and the advent of highly sophisticated ultrastructural studies, it became clear that the flagellates are the most primitive representatives of the Protozoa. The view that the flagellated Protozoa are the most primitive and that the Sarcodina are closely related and derived from them is now widely accepted. However, debate still continues. Sleigh (1979) argues that the amoebae are polyphyletic in origin, suggesting that some were derived from chrysophyte flagellates or their ancestors, while others may have originated at one or several much earlier stages during the development of eukaryotes. This view is now widely accepted. Certainly our knowledge of the morphology and physiology of the amoebae, particularly the naked group, is at present limited, largely because of the difficulties of identification and lack of techniques, so that there is considerable difficulty in studying the phylogeny of these Protozoa.

The phylum Ciliophora may well represent one of the most homogeneous groups within the Protista. They are a large compact group which are readily distinguishable from all the other protozoan groups. There are a series of major diagnostic features held in common by almost all members of the Ciliophora. Almost without exception ciliates exhibit nuclear dualism, possessing one or more diploid (occasionally polyploid) micronuclei and one or more polyploid or polygenomic macronuclei. The organelles used for locomotion and feeding are cilia or compound ciliary structures. A cell mouth or cytostome is commonly present. There are however some exceptions; among the Apostomatidea there are astomatous species and the Suctoria are polystomatous in possessing numerous feeding tentacles rather than a single oral opening. There is an absence of true syngamy and the mode of binary fission is homothetogenic (in contrast to symmetotrogenic binary fission exhibited by other protozoan groups, especially flagellates) (Corliss, 1979).

Cilia and flagella are structurally similar, and it is suggested that ciliates may be derived from a zooflagellate ancestor. Although this cannot be substantiated with hard evidence, Corliss (1956, 1960), who has considered the evolution and systematics of the ciliates in detail, considers a zooflagellate ancestory a reasonable hypothesis. Among the ciliates the hypotrichs are considered the pinnacle of protozoan evolutionary development.

D. Free-living Protozoa

From Table 1.1 it is clear that many of the protozoan groups are exclusively parasitic. The free-living members of the Protozoa are members of the superphyla Mastigophora and Sarcodina and the phylum Ciliophora. The majority of flagellates, sarcodines and ciliates are free-living, but each group contains some species which have adopted a parasitic mode of life. Indeed among the flagellates and amoebae are parasitic species of some medical importance. Various forms of trypanosomiasis, among them sleeping sickness, are caused by flagellates, and amoebic dysentry is caused by *Entamoeba histolytica*.

All the free-living Protozoa are essentially aquatic, living in the benthic and planktonic communities of freshwater, brackish and marine environments. Many live in semi-terrestrial habitats, in damp moss beds or in the water films around soil particles. The Protozoa are ubiquitous and cosmopolitan in their distribution worldwide. The species that live in arctic seas, for example, are also found in temperate waters. While some species thrive in the water and soils of the world's polar regions, others have adapted to high temperature environments. There are numerous reports of protozoans living in hot springs at extremely high temperatures, for example *Chilodon* at 68°C (Dombrowski, 1961) and *Oxytricha fallax* at 56°C (Uyemura, 1936). However, as Tansley and Brock (1978) point out, such reports must be substantiated with experimental cultures at the recorded temperatures. The fact that an organism can survive at an extreme temperature does not imply that it functions and grows normally at that temperature.

Even the man-made environment of sewage treatment plants, both activated sludge and filter-bed processes, have been colonised successfully by Protozoa. Indeed ciliates have been shown to perform an important role in the production of good quality effluents in the sewage treatment process (Curds and Cockburn, 1970a, 1970b).

(i) Mastigophora – The Flagellates

This is a particularly heterogeneous group, both structurally and physiologically. Typically all members of this superclass have one or more flagella which as act locomotory and feeding organelles. Only one type of nucleus occurs and usually cells are uninucleate, but some species are multinucleate. The Mastigophora are traditionally divided into the Phytomastigophorea and Zoomastigophorea (Table 1.1). The former contains those taxa of flagellates of which the majority bear chloroplasts and hence are wholly or partially autotrophic. Chlorophyll a is the major photosynthetic pigment, but other accessory pigments including chlorophylls b and c, carotene and xanthophyll also occur in flagellates. The structure and number of the chloroplasts in the various groups of Phytomastigophora are variable and in many species a pyrenoid is present in the chromatophore. The pyrenoid is a dense area associated with polysaccharide formation. A small conspicuous stigma or eyespot positioned at the anterior end of the cell is characteristic of many species. These structures, which are made up of a cluster of lipid globules containing carotenoid pigments, are involved in orientation towards light and phototaxis. Many of the autotrophic protozoans live as colonial entities, e.g. *Volvox* (Figure 1.1), with many cells embedded

Figure 1.1: *Volvox*, a Colonial Green Flagellate, Containing Numerous Individual Cells in a Mucilagenous Matrix. The large bodies shown in the photograph are developing daughter colonies (x 480).

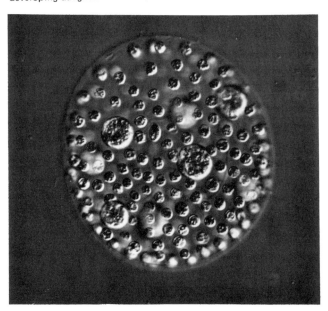

Figure 1.2: *Euglena gracilis* (x 1,250).

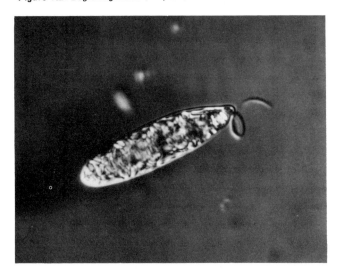

in a mucilagenous matrix, while others, e.g. *Euglena* (Figure 1.2), function as single cells.

Figure 1.3: The Collared Flagellate *Diplosiga socialis*.

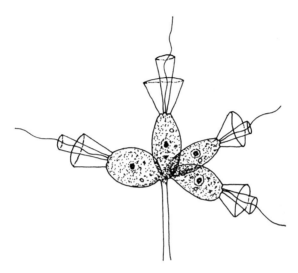

Zoomastigophorea are heterotrophic and may live as single cells, or like many of the Phytomastigophorea as members of a colony (Figure 1.3). In both groups binary fission is symmetrigenic, occurring in a longitudinal plane. Sexual reproduction is not a widely-reported characteristic, but when present is essentially syngamy.

(ii) Sarcodina – The Amoebae

Typically members of the Sarcodina possess pseudopodial structures which are used for movement and feeding. Such structures show considerable diversity within the various groups of sarcodines. Morphologically amoebae fall into two broad categories, the naked amoebae and the testate or shelled amoebae. The former, as the term implies, lack any form of skeletal structure (Figure 1.4), while the latter often possess elaborate shells or tests which may be proteinaceous, agglutinate, siliceous or calcareous in composition and are usually constructed as a single chamber with a single aperture, although in the foraminiferans shells with numerous chambers are common (Figures 1.5 and 1.6). Other supportive structures occur in the Actinopoda which possess an array of axopodia (Figure 1.7), each of which is supported by a central fibrous axis composed of microtubules. These microtubular

Figure 1.4: *Amoeba proteus* under Nomarski Interference Microscopy (x 420). n — nucleus, cv — contractile vacuole, dv — digestive vacuole.

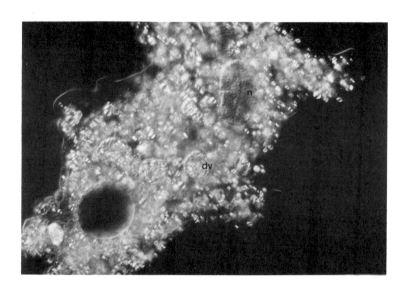

Figure 1.5: Foraminiferan Shells. a — *Globigerina bulloides*, b — *Asterigerina carinata*.

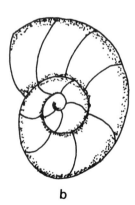

a   b

Figure 1.6: The Test of *Arcella vulgaris* Viewed from Below. Inset shows the organism in side view. (Nomarski interference microscopy x 450 and x 1,250.)

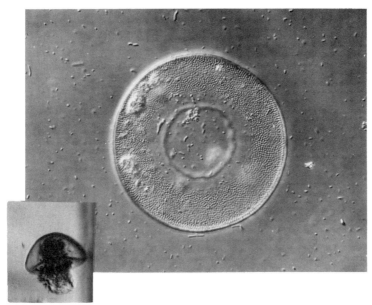

Figure 1.7: *Actinosphaerium* under Nomarski Interference Microscopy (x 1,250).

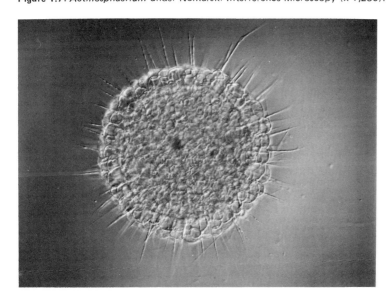

structures, the axonemes, extend within the body of the cell where they originate from the cortical cytoplasm, a central granule or nuclear membrane. Members of this superclass may also have mineral skeletal structures composed of organic matter, silica or strontium sulphate.

The cytoplasm of amoebae is differentiated into the fluid endoplasm which is granuolated or vacuolated in appearance and contains the nucleus, food vacuoles, contractile vacuoles and other inclusions. The surrounding stiffer ectoplasm is hyaline and appears homogeneous in structure. Usually amoebae are uninucleate, but some multinucleate species do occur. Nuclear dualism is a feature of the developmental stages of some species of Foraminiferida, a characteristic these protozoans hold in common with ciliates, although the phenomenon developed independently in each group (Grell, 1974). In shelled species the cytoplasm usually fills the chamber in small species such as *Euglypha*, but in larger species, for example *Difflugia*, the cytoplasm only partially occupies the chamber and thin cytoplasmic strands attach it to the shell wall. Asexual reproduction by binary fission in an undefined plane is usually the rule in naked forms, but in testate and shelled amoebae fission may be longitudinal or transverse, or multiple fission or budding may be practised among the various species in these groups. Sexual reproduction is rare and if present is associated with flagellate and occasionally amoeboid gametes. In the foraminiferans alternation of sexual and asexual generations is common.

(iii) Ciliophora – The Ciliates

This is a large homogeneous group, in which over 7,000 species have been described (Corliss, 1979); no doubt many more await discovery. The ciliates are characterised by their complex ciliated cortex, where simple cilia or compound ciliary structures or cirri are arranged over the cell in an ordered fashion and serve the function of effecting locomotion. Cilia and ciliary structures associated with the cytostome facilitate feeding, and produce feeding currents in those species which are filter feeders. The gymnostomes mostly lack oral ciliature, and frequently the cytopharynx is strengthened by rods or trichites (Curds et al., 1983). These ciliates are essentially macrophagous in their feeding behaviour.

Ciliates characteristically show nuclear dimorphism, usually possessing one macronucleus concerned with the regulation of normal cell functioning and one or more micronuclei which are responsible for the replication of genetic material during reproduction. Macronuclei are very diverse morphologically, and may be spherical, oval, ribbon-shaped, horse-shoe-shaped or longitudinally stretched with constrictions separating the long structure into nodes, as shown in Figure 1.8. As a general rule large ciliates tend to have large macronuclei.

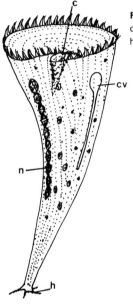

Figure 1.8: *Stentor*. c — cytostome, cv — contractile vacuole, n — nucleus, h — holdfast.

The evolution of nuclear dimorphism in the ciliates obviously occurred earlier than the polyploidy which is a characteristic of macronuclei, and went through a series of stages of which some evolutionary phases have representatives which have survived as relics (Raikov, 1969). In the early stage, some characteristics of the ciliate group were present but nuclear dimorphism was not. The second stage is characterised by the development of nuclear dualism with genetically identical nuclei differentiated into somatic (macronuclei) and generative (micronuclei) nuclear types. At this stage the macronuclei are diploid and have no ability to divide and thus degenerate at reproduction. Ciliates which belong to this group are the Loxodidae and the Geleiidae. The final evolutionary stage was reached when polyploidism of the macronucleus was achieved, and polyploid macronuclei occur in the majority of ciliates. The development of polyploidy must have evolved more or less simultaneously with a reacquired ability for macronuclear division.

Ciliates reproduce asexually by homothetogenic binary fission, which typically involves a plane of division perpendicular to the antero-posterior axis of the body, resulting in perkinetal fission. This is commonly called transverse fission. There are of course some exceptions to this rule, but in all cases stomatogenesis occurs at some stage during division. The suctorian Ciliophora do not possess cilia in what may loosely be called the 'adult' sessile stage. These organisms are attached by non-contractile stalks to substrata of various types, and possess feeding tentacles which usually capture and feed on other Protozoa. The Suctoria reproduce asexually by a process of endogenous or exogenous budding. The young thus produced are ciliated and then disperse, undergoing transformation into the sessile 'adult' stage on settling, by loss of the cilia and the development of a stalk and tentacles. The ability to carry out a form of sexual reproduction is common among ciliates, and usually occurs by conjugation or autogamy. Total conjugation occurs in some groups, and has been regarded, misleadingly, as true syngamy by some protozoologists. The sexual phase does not result in an increase in individuals and is essentially carried out to achieve genetic exchange. Some examples of ciliates are shown in Figures 1.8 and 1.9.

E. The Protozoan Cell

Each protozoan cell functions as an independent entity and carries out all the essential life processes that occur in metazoan organisms, albeit at a less sophisticated level. Instead of tissues or organs, we find organelles developed to carry out the fundamental functions of life. Free-living Protozoa are capable of locomotion, often moving extremely fast in relation to their size, by means of flagella, cilia or pseudopodia. Among the heterotrophic Protozoa there is a wide diversity of trophic type; bacterial feeders, algal feeders, carnivores feeding on other protozoans and omnivorous feeders exploiting several trophic levels. There is diversity not only in trophic status, but also in the manner in which food is collected or captured before it makes its way into the food vacuoles of the cell. Most ciliates have well-developed cell mouths so that food enters at a definite point and waste material leaves by the cell anus or cytopyge, whereas other Protozoa, such as the amoebae, lack a mouth and food therefore has no fixed point of entry or exit from the cell. Yet even in what are seemingly simple amoebae there are a range of ways in which the feeding process occurs. Protozoa have been shown to exercise selectivity in feeding (Schaeffer, 1910; Seravin and Orlovskaja, 1977), and are able to distinguish palatable and non-palatable bacteria and items of prey. Some species seem to have their own limited range of preferred food organisms.

Figure 1.9: a: *Paramecium caudatum* (x 450). b. *Vorticella microstoma* (x 500).

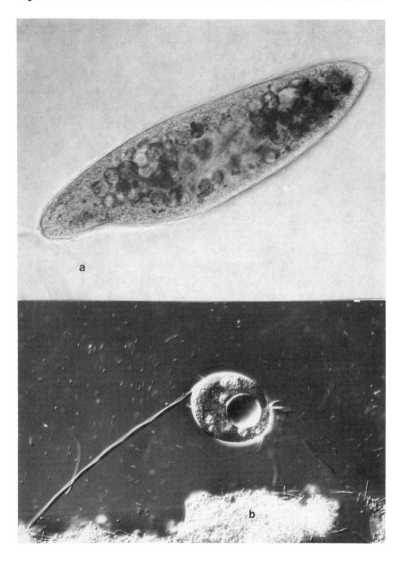

The various modes of asexual reproduction are amazingly diverse, and wide variations can be achieved on the basic theme of binary fission. Protozoan cells increase in size during growth, and when they attain a particular size, the cell divides into two, usually equal halves. The plane of division and the sequence of events varies among the Protozoa. In some species binary fission has evolved into budding or multiple fission. The size reached by a particular species before division occurs varies in response to factors such as temperature and food supply. Sexual reproduction is not a universal characteristic of Protozoa. In those species that possess sexual competence the sexual form of reproduction is usually only resorted to under conditions of adversity, when food supply becomes depleted, or when the physical and chemical environment becomes unfavourable. The normal alternation of sexual and asexual generations in the foraminiferans is exceptional among free-living Protozoa.

Small organisms depend upon diffusion for the acquisition of oxygen and the passage out of the cell of some waste products of metabolism. There are reports of haemoglobins in some ciliates, but these have an extremely high affinity for oxygen and do not function during normal respiratory processes (Ryley, 1967). The size a protozoan can reach is probably governed by the dependence on diffusion. The larger fixed-shape species tend to be long and thin, thus faciliating the passage of oxygen to all parts of the cell, while the larger species of amoebae have a large surface area by virtue of the pseudopodia they extrude, and a continuously changing shape, thus all parts of the cell will have access to the oxygen in the environment by diffusion. A comparative consideration of the respiratory physiology of large and small ciliates suggests that species the size of *Stentor* may be near the size limit which can be achieved in an independent one-celled organism.

While naturally there are limitations on what can be achieved at a single-cell level, nonetheless the Protozoa during the course of evolution have exploited most of the conceivable possibilities. Being small and single-celled clearly has some positive advantages. Rapid asexual reproduction allows the possibility of mutation producing individuals better able to cope with changed conditions. Presumably the widespread ability of Protozoa to encyst developed early in their evolutionary history. This characteristic allows survival for long or short periods under adverse conditions, and widespread dispersal by the wind when aquatic habitats or semi-terrestrial situations dry up. A simple physiology is more readily adaptable than a complex specialised one. Despite a small size and a relatively simple form, the Protozoa are unarguably a successful group. There are an enormous number of species; at the time of the last revision of protozoan classification (Levine *et al.*, 1980) there were 65,000 named species, of which over half are fossil and about 10,000 parasitic. Free-living species have colonised an amazing diversity of habitats, including many hazardous and extreme environments, while parasitic species have in many cases developed complex life-cycles involving several hosts. Other species have opted for a symbiotic role in the digestive tracts of vertebrate and invertebrate herbivores where they perform a critical part in the breakdown of cellulose.

While the study of protozoan ultrastructure and biochemistry has forged ahead, elucidating the mechanisms by which Protozoa feed, move and perform other physiological functions, their ecological energetics and role in the communities and ecosystems they inhabit have, until relatively recently, been overlooked. Indeed their small size may belie their function in many habitats. Many are part of the saprovore food web, exploiting the microorganisms which bring about decomposition and the recycling of minerals. Evidence is now beginning to accumulate which suggests that Protozoa, by their grazing activities, and possibly also by the secretion of growth-promoting substances, may stimulate some decomposer microorganisms, thus enhancing the essential process of decomposition and nutrient recycling.

2 HOW PROTOZOA OBTAIN ENERGY

A. Introduction

Protozoa have been classified into three trophic categories: the photo-autotrophs which harness the sun's radiant energy in the process of photosynthesis; the photoheterotrophs, which although phototrophic in energy requirements, are unable to use carbon dioxide for cell synthesis and must have organic carbon compounds; and lastly the chemoheterotrophs which require chemical energy and organic carbon sources (Nisbet, 1984). The latter group, which are more commonly referred to as heterotrophs, include the majority of Protozoa, while the first two trophic categories are largely restricted to members of the Phytomastigophorea.

Autotrophic flagellates are able to synthesise carbohydrates from carbon dioxide and water by chlorophyll using the radiant energy of the sun and converting it to chemical energy in the following way:

$$6H_2O + 6CO_2 \, (+\,674 \text{ kcal}) \xrightarrow{\text{chlorophyll}} C_6H_{12}O_6 + 6O_2$$

The reaction is reversed in respiration, which usually proceeds at a slower rate, so that a net gain of organic matter, referred to as primary production in ecological terms, is produced by all chlorophyll-bearing organisms and forms the major part of the first step in the transfer of the sun's radiant energy through the biological world. The organic matter, or potential energy, is exploited progressively through herbivory and carnivory by heterotrophic organisms. Any energy which is not lost as heat during respiration in this food chain eventually makes its way as unconsumed primary production, corpses or faeces into the pool of dead organic matter, which in turn is exploited as an energy source by the decomposer microorganisms and detritivores. The microorganisms in turn are consumed by microbivore animals, and these together with detritivores are predated by carnivores.

Many of the photoautotrophic flagellates, largely members of the Euglenida, Cryptomonadida and Volvocida combine autotrophy with heterotrophy in varying degrees, and are often described as the acetate flagellates, their preferred carbon sources being acetates, simple fatty acids and alcohols. These flagellates are able to switch from autotrophy in the light to heterotrophy in the dark, providing the required substrate is available in the medium. Within these orders obligative autotrophy and facultative heterotrophy may be found in closely-related species. *Euglena pisciformis*, for example, cannot survive if maintained in the dark on organic media and is entirely dependent on autotrophy, while *Euglena gracilis* can utilise organic media when deprived of illumination.

The majority of free-living Protozoa are heterotrophic, exploiting a wide range of diet and consequently occupying a number of trophic levels. Some species feed on bacteria and are therefore microbivores in the decomposer food chain, while others feed on algae, usually the unicellular variety, and are thus herbivores. Both trophic groups are exploited by carnivorous Protozoa, many of which also feed on other micro- and meiofauna such as rotifers, gastrotrichs and small crustaceans. Within the Protozoa we find a spectrum of trophic types — autotrophs, primary consumers and secondary consumers all inter-related in a complex community food web.

As far as feeding is concerned, the free-living Protozoa can be divided on morphological grounds into two groups: those with a mouth or cytostome and those lacking a mouth or definite point of entry for food. Some of the flagellates and almost all of the ciliates, with the exception of some members of the Apostomatida, possess a cytostome. The suctorian Ciliophora do not exhibit the typical ciliate cytostome, here each feeding tentacle being essentially a mouth. The Sarcodina and many of the flagellates have no mouth, although in each group there are some species which take food into the cell through a particular region on the cell surface. The quantity of energy ingested and very often the feeding behaviour, are subject to modification by biological and environmental factors in all free-living Protozoa.

B. Modes of Feeding

(i) Protozoa with a Cytostome or Cell Mouth

The cytostome is the usual form of mouth encountered among ciliates and some flagellates, but the tentacles of the suctorians are also cell mouths of a very different structure. The cytostomes of ciliates show a progression from the relatively simple to the complex. The gymnostomes show a simple oral structure in which most species are without oral ciliature, and the cytostome is at the surface of the body in an apical or lateral position and leads into a well-developed cytopharynx supported by bundles of microtubules or nematodesmata. In the Hymenostomatia there is a definite oral ciliature composed of three or four specialised membranelles situated in the buccal cavity (Figure 2.1). Usually the adoral zone of membranelles is on the left of the buccal cavity while on the right there is a single paroral membrane. The peritrichs have highly specialised oral ciliature which winds anti-clockwise down into an infundibulum towards the cytostome (Figure 2.2). The spirotrichs are characterised by a well-developed conspicuous adoral zone of many membranelles which may extend out onto the surface of the cell. These membranelles are used for both feeding and locomotion. The most complex ciliates are the hypotrichs in which the ventral cilia may be formed into cirri which are arranged in rows or groups. The buccal cavity is usually in the anterior region of the ventral surface. The ciliature may include a well-developed undulating membrane in addition to a prominent zone of adoral membranelles which usually extend along the anterior edge of the cell (Figure 2.3).

The oral region of ciliates has been the subject of much ultrastructural research, the majority of which has aimed at elucidating the processes involved in concentrating and capturing food material, the formation of food vacuoles and aspects of the digestive process. Unfortunately there is considerable variation in the terminology applied to particular ultrastructural characteristics, which can result in difficulties of interpretation. In addition to the external components of the buccal opening, the buccal tube and the cytopharyngeal-cytostomal region exhibit characteristic structures. The membranelles often give rise to the microtubular nematodesmata which are common components of the buccal apparatus of many ciliates (Piltelka, 1968; Fischer-Defoy and Hausmann, 1981). Their role is not clear, but they may act as anchorage for membranelles or contribute to the transport of food into the cell. Distributed among the nematodesmata are elongated vesicles termed variously 'coated cisternae' in *Climacostomum virens* (Fischer-Defoy and Hausmann, 1981), 'cisternae' in *Tetrahymena* (Allen, 1967) and 'membranous sacs' in *Paramecium putrinum* (Patterson, 1978). Their function is uncertain, but their location and their appearance distinguish them from the pharyngeal or discoidal vesicles found associated with the cytopharynx.

Cytopharyngeal bands or ribbons of microtubules have been described in a number of ciliates including *Paramecium* (Allen, 1974) and

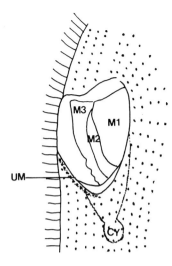

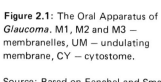

Figure 2.1: The Oral Apparatus of *Glaucoma*. M1, M2 and M3 — membranelles, UM — undulating membrane, CY — cytostome.

Source: Based on Fenchel and Small (1980).

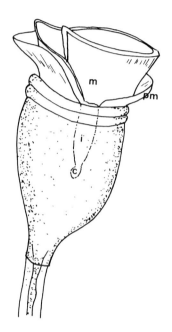

Figure 2.2: The Oral Apparatus of a Peritrich. pm — paroral membrane, m — membranelle, i — infundibulum, c — cytostome.

Figure 2.3: *Euplotes*, Showing Cirri (c) and the Adoral Membranelles (am). oa — oral aperture.

Climacostomum virens (Fischer-Defoy and Hausmann, 1981) — Figures 2.4 and 2.5 — but their pattern of distribution varies among those ciliates so far studied. In *Paramecium*, for example, the cytopharyngeal ribbons arise in a dense filamentous reticulum associated with the buccal apparatus. In *Climacostomum*, on the other hand, they arise from the circular haplokinety. Haplokinety is the term applied to the double row of kinetosomes (of which one row is barren) from which the paroral membrane arises. Associated with the microtubular ribbons are discoidal pharyngeal vesicles (Figure 2.4). These vesicles vary in appearance among ciliates and have been variously termed. The discoidal vesicles perform a function in providing membrane material in the formation of food vacuoles (Allen, 1974; Kloetzel, 1974; Fischer-Defoy and Hausmann, 1982). A cytostomal cord has been described in a number of species including *Climacostomum* (see Figure 2.4), but its function is unclear.

During the feeding process, food which is drawn into the cytostomal-cytopharyngeal region is enclosed in food vacuoles. The food vacuole develops by the fusion of cytopharyngeal membrane and cytopharyngeal discoidal vesicles. The series of events following this process are outlined in Section 2 D. When Protozoa are feeding rapidly they produce large amounts of surface plasma membrane, which is utilised in the formation of food vacuoles. Studies on the rate of digestive or food vacuole formation suggest that the membrane of vacuoles equivalent to 50-150 per cent of the total cell surface is produced in 5-10 minutes in a species like *Euplotes* (Kloetzel, 1974). In many ciliates the membrane components are recycled in the cell. Studies on *Paramecium* have revealed an elaborate system of microtubular ribbons which arise at the left side of the cytopharynx and fan out into the cytoplasm, where some pass to the cell anus or cytoproct (Allen, 1974). Membrane is retrieved at the cytoproct and moved along the microtubular ribbons to the cytopharynx, where it enters the discoidal vesicle pool (Allen and Fok, 1980). In *Climacostomum* the membrane of defaecation vacuoles has been seen to be retained in the cell and fragments to form vesicles which are presumably recycled to the cytopharynx (Fischer-Defoy and Hausmann, 1982). It should be noted however that this process may not occur in the same manner in all ciliates; indeed Hausmann and Hausmann (1981) were unable to find microtubular structures associated with guiding discoid vesicles from the cytoplasm to the cytostome in the peritrich *Trichodina pediculus*, and the morphological features of the vesicles in this species suggest that the vesicles may not be recycled directly from defaecation vacuoles. Clearly much more work is needed

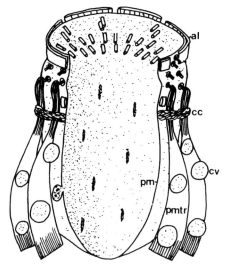

Figure 2.4: The Cytostome-Cytopharynx Region of *Climacostomum virens*. al — alveolus, cc — cytostomal cord, cv — cytopharyngeal vesicle, pm — plasma membrane, pmtr — post-ciliary microtubular ribbon.

Source: Fischer-Defoy and Hausmann (1981), with the permission of Springer-Verlag, Heidelberg.

Figure 2.5: The Buccal Cavity of *Paramecium* as Viewed from Inside the Cell. cr — cytopharyngeal ribbons, cy — cytostome-cytopharyngeal complex, fv — position of developing food vacuole, q — quadrulus, rw — non-ciliated ribbed wall, v — vestibule.

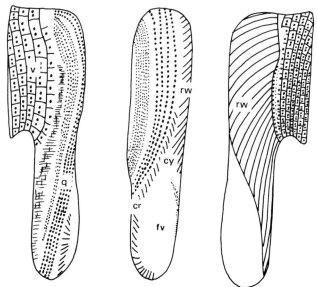

Source: Redrawn from models constructed by Allen (1974).

in this interesting area on a much wider range of species.

In almost all species so far studied the food vacuoles progress through the cytoplasm during a digestive cycle, and in ciliates eventually end up at the cell anus or cytoproct, where undigested material is voided from the cell. In sarcodines and flagellates this structure is absent and waste material is voided at random points on the cell surface. In the ciliate *Stylonychia mytilus*, however, there appears to be an intracellular system of digestive channels. *S. mytilus* is capable of ingesting and accommodating large numbers of food organisms, including bacteria, flagellates and other ciliates. The cytoplasm has two distinct regions, an area of membrane-bound cytoplasm containing the cytoplasmic organelles and an area of large spaces or channels with coiled tubular bodies dispersed in it. These channels are a permanent feature of the cells in a wide range of physiological states. The channels are occupied by prey undergoing digestion and the food organisms appear to be devoid of any limiting membrane. The limiting membrane of the ground cytoplasm separates prey from the main cytoplasm. When the flagellate *Chlorogonium* is preyed upon, however, it is found in the cytoplasm enclosed in a limiting membrane for about 30 minutes; thereafter it moves to the channels where the limiting membrane disappears (Dass *et al.*, 1982). It is suggested that the presence of such digestive channels allows high rates of ingestion and does away with the need for recycling vacuole membrane material and the formation of food vacuoles.

Ciliates vary in the manner in which they collect or capture their food as a function of diet and the structure of the oral apparatus. The majority of species exploiting small particulate food, such as bacteria, small flagellates and ciliates, are essentially filter feeders, creating feeding currents by means of the oral ciliary structures and collecting and concentrating particles in the current. The ability to clear water and concentrate the available food material in it, is considerable in ciliated Protozoa. In *Tetrahymena pyriformis* and *T. vorax* a volume equivalent to the buccal cavity is cleared in less than one second and particles in the feeding current are concentrated by 500 times or more (Rasmussen *et al.*, 1975). In a recent study Fenchel (1980a) showed that bacterivore holotrichs can typically clear 3×10^3 to 3×10^4 times their own volume of water in an hour. For a ciliate the size of *Tetrahymena* this represents about 5×10^{-5} ml h^{-1}. Ciliates the size of *Paramecium*, *Euplotes* or *Blepharisma* feeding on large bacteria, yeasts and microflagellates clear between 2×10^{-4} and 2×10^{-3} ml h^{-1}. It appears that each species has a distinct size spectrum of particle within which it can retain and ingest successfully, and this is a function of the mouth morphology (Fenchel, 1980b). Most bacterivorous holotrichs can retain particles down to 0.2 μm, retaining particles between 0.3 and 1.0 μm most efficiently. The spirotrichs investigated did not retain particles smaller than 1-2 μm. No selectivity appears to be practised in the type of particle retained; it is simply the size of the particle rather than its suitability as food or its energy value which dictates whether or not it is ingested. Furthermore, the rate of water transport is constant and independent of particle concentration in the water. Thus unlike many metazoan filter feeders, the Protozoa do not appear to modulate filtering activity in relation to particle density and palatability.

The generation of feeding currents in a wide range of ciliates, 17 species in all, feeding on latex particles has been investigated in considerable detail by Fenchel (1980c). In higher ciliates, the Oligohymenophora and Polyhymenophora, the water currents are generated by ciliary membranelles which are situated on the left side of the cytostome and propel water in a direction almost parallel to themselves. In most cases the membranelles form three parallel rows of dense cilia which generate metachronal waves. In the oligohymenophores (hymenostomes and peritrichs) the membranelles are situated in the bottom and along the left margin of the buccal cavity and direct water posteriorly towards the cytostome. The water currents are forced upwards and anteriorly and are intercepted by the paroral membrane where particles in the water are trapped in the free spaces between the cilia, thus concentrating the food material between the paroral membrane and cytostome. The pattern of flow in peritrichs (Figure 2.6), has been described by Sleigh and Barlow (1976) and by Fenchel (1980c). Observed from the oral pole, the water moves in an anti-clockwise direction down into the infundibulum, drawn along by metachronal waves generated by complex series of membranelles. Peritrichs have a much larger filtering area on the paroral membrane by virtue of the oral torsion they possess. Among the polyhymenophorans (spirotrichs) the paroral membrane may be completely absent, but if present it plays no functional part in food collection. In this group of ciliates the membranelles not only generate the feeding currents, but also act as a filter. Water is pumped out of the buccal cavity by the membranelles and particles which cannot pass between two adjoining membranelles are retained in the buccal cavity. The retained particles are washed posteriorly along the membranellar zone and are concentrated at the cytostome. Thus in most ciliates studied, with the exception of the polyhymenophorans, the membranelles generate the currents and the paroral membrane acts as the filter. However, among the hymenostomes *Glaucoma chattoni* and *G. scintillans* do not conform to the general pattern because in these species it is the third membranelle (see Figure 2.1), not the paroral membrane, which intercepts the food particles (Fenchel and Small, 1980).

Figure 2.6: The Feeding Currents of a Peritrich. A: the whole organism indicating the position of the infundibulum (i), cytostome (c) and paroral ciliary structures (pc). B: the infundibulum and the direction of feeding currents.

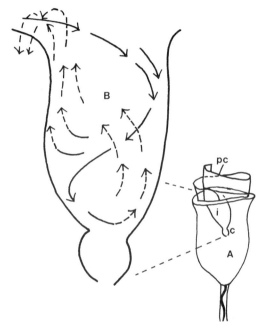

The ciliary feeding mechanism in *Paramecium*, and some other ciliates feeding on small items, appears to involve a mucoid secretion (Jahn *et al.*, 1961). Particles of food are agglutinated as they are driven along the oral groove by ciliary beating. The mucus is supplied by peristomal sacs at the bases of the oral groove cilia. The secretion of mucus by ciliates has been noted by several researchers (Hardin, 1943; Curds, 1963). The secreted mucus may cause the flocculation of bacteria in the water or medium. Ciliates are often seen actively feeding around such floccules in cultures, presumably encountering higher capture rates of bacteria in the feeding currents produced in the vicinity of dense bacterial colonies. The secretion of mucus may play a dual role in the feeding behaviour of many ciliates; first it aids in drawing food into the cytostome and secondly it concentrates bacteria in the medium, making them easily accessible to grazing.

Figure 2.7: *Didinium* Feeding on *Paramecium*. e — extrusomes, s — seizing organ.

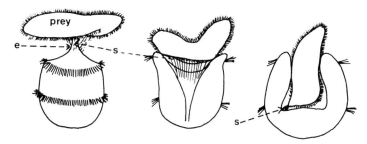

Source: Redrawn from Mast (1906).

Many carnivorous species prey on large organisms, often as big as or larger than themselves. Frequently the prey are ingested whole. Some means of capturing and immobilising the prey is necessary, after it has been recognised as a suitable food organism. There is evidence that chemoreception plays a role in the food selection of species such as *Didinium nasutum*, *Dileptus anser* and *Lacrymaria olor* (Sevarin and Orlovskaja, 1977). The apprehending mechanism involves various types of extrusomes, notably toxicysts, which are found in the vicinity of the oral region and contain toxins which paralyse the prey. In *Didinium* a second form of extrusome occurs. Termed pexicysts, they are short attachment rods lying below the surface of the oral cone and function in attaching to the surface of the prey, while the toxicysts penetrate into the prey cytoplasm (Wessenberg and Antipa, 1970). *Didinium* has a very limited repertoire of prey, feeding largely on members of the genus *Paramecium*, although it has been successfully reared experimentally on other ciliate species (Berger, 1979) — Figure 2.7.

Traction or suction-like processes may be practised by ciliates which feed on filamentous algae. Such species include *Frontonia*, *Nassula* and *Chilodonella*. A particularly elegant study of the process of ingestion and the structures facilitating ingestion, has been carried out by Tucker (1968) on *Nassula*. Like all members of the class Kinetofragminophorea, in which the carnivorous forms previously mentioned are also included, *Nassula* lacks compound oral ciliature. *Nassula* possess a type of cytopharynx which is commonly referred to as a cytopharyngeal

Figure 2.8: The Cytopharyngeal Basket of *Nassula*.

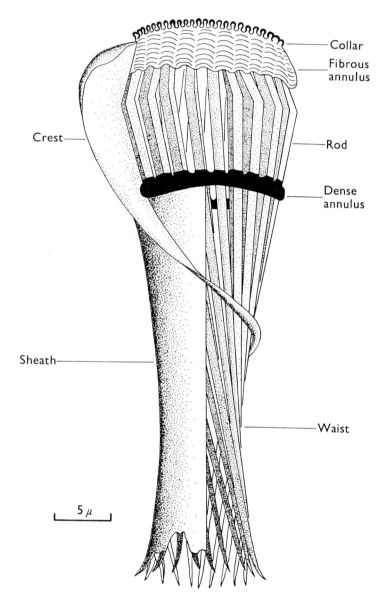

Source: Tucker (1968), with permission of the Company of Biologists Ltd.

basket. As shown in Figure 2.8, the cytopharyngeal basket is made up of cytopharyngeal rods forming a circular pallisade surrounded by a dense annulus, and below this by a sheath. The top of the basket is encircled by a fibrous annulus. All of these components are made up of microtubules. The feeding process is illustrated in Figure 2.9. During feeding *Nassula* positions itself with the top of the basket alongside an algal filament. The top of the basket is then closely applied to the filament and the top of the pallisade becomes elliptical in cross-section as the tops and bottoms of the rods bordering the longer side of this ellipse become widely spaced (Figure 2.9ii). A hemispherical hyaline extrusion bulges out of the cytostome and engulfs the algal filament which bends into a hair-pin shape as it moves into the lumen of the cytopharynx. At this point the rods become widely spaced near the mid-level of the pallisade which dilates (Figure 2.9iii). Once the bent portion of the filament has passed through the lumen, the pallisade again becomes circular in cross-section and constricts at all levels, pressing the two strands of the alga together (Figure 2.9iv). Finally the pallisade dilates at all levels above the dense annulus and constricts all the way down below it as the rods return to the configuration shown in Figure 2.9i. This process of feeding can be rapid; *Nassula* is capable of ingesting a length of algal filament 1.6 mm long (eight times its own length) in a matter of 4-5 minutes. Shorter lengths of algae enter more rapidly. The algal cells separate after about three minutes once inside *Nassula*. Tucker (1968) suggests that enzymes stored in cytopharyngeal vesicles may be responsible for the break-up of the algal filaments.

A number of flagellates also possess a cytostome. However, this group has attracted less attention than the more sophisticated ciliates, despite the fact that from a nutritional point of view the Mastigophora probably show much wider diversity of feeding and nutritional types than any other group of Protozoa (Nisbet, 1984). Some members of the euglenoid group of flagellates feed on particulate matter via a permanent cytostome. *Peranema* is probably the best-known example of this type of flagellate. Associated with the cytostome are two flagella and a rod-like structure, the rodorgan. The rodorgan is composed of a pair of rods, each constructed from 100-200 microtubules, enclosed in a sheathing membrane anteriorly (Nisbet, 1974). *Peranema* feeds on a variety of food organisms, including bacteria and detritus, but can also successfully ingest food items as large as itself. When feeding on *Euglena*, for example, the rodorgan and part of the adjacent region of the ventral body surface move forwards until it touches the prey. The rodorgan protrudes and becomes attached to *Euglena*, while the body of *Peranema* then moves forwards and the prey moves into the expanded cytostome. The rodorgan detaches from the prey, moves over its surface and reattaches at another point, while the body of *Peranema* continues to move forwards. Thus ingestion is achieved by repeating attaching, pushing, detaching and reattaching of the rodorgan. The whole process requires 2-15 minutes (Chen, 1950). The movement of the rodorgan is performed by an elaborate arrangement of articulating lamellae connected to the bases of the pair of rods in the organ (Nisbet, 1984).

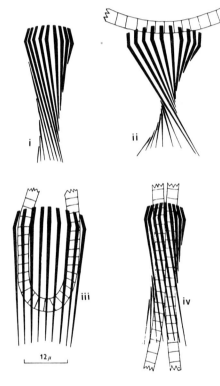

Figure 2.9: The Process of Ingesting a Filamentous Alga in *Nassula*.
Source: Tucker (1968), with the permission of the Company of Biologists Ltd.

The Suctoria are a highly specialised group of ciliate predators. The 'adults' are sessile; the cilia are lost and replaced by an array of feeding tentacles, which in some species may be distributed over the whole cell surface and in others arise from localised areas of the cell. The suctorians are essentially polystomate, each feeding tentacle representing a cell mouth. Tentacles are usually of two functional types, suctorian tentacles for feeding, and piercing tentacles concerned with capturing and immobilising the prey. When feeding, the tentacles are outstretched in order to maximise the chances of prey coming into chance collision with a tentacle. Ciliates are the usual prey and when a suitable ciliate makes contact with a tentacle it is immediately captured. As the prey struggles, other tentacles orientate onto the prey surface and become attached. After a few minutes the captured ciliate, which is often larger than its suctorian captor, ceases struggling. A single suctorian can feed simultaneously on a number of prey (Figure 2.10).

Figure 2.10: *Podophrya* Feeding on *Colpidium*. In this case five prey are being fed upon simultaneously.

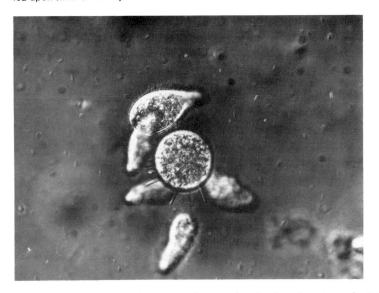

The mechanism of adhesion to the prey has intrigued protozoologists for many years. Hull (1961a) suggested that an adhesive interaction occurs between a secreted substance on the pellicle of the prey and another substance on the suctorian tentacle. The fact that adherence is enhanced by divalent ions in the medium in catalytic amounts and by the presence of sulphhydryl compounds, and is inhibited by low temperatures, justifies this hypothesis and is suggestive of an enzyme-catalysed reaction. The evidence suggests that acetylcholine is one of the more important substrates in the ciliate pellicle for the reaction. It is envisaged that divalent ions and sulphhydryl compounds have a role as cofactors in the reaction, the sulphhydryls acting as hydrogen transport substrates.

The mechanism by which the Suctoria transport the organelles and cytoplasm of their prey along the tentacles into food vacuoles has generated a great deal of ultrastructural research on the tentacle. Tucker (1974) has described the structure of the tentacle in *Tokophrya* in detail. Microtubular arrays run along the entire length of each tentacle and project for several micrometres into the cell body beyond the base of the tentacle. A knob is situated at the end of each tentacle. Within the tentacle there are seven microtubular rows surrounded by outer tubules encircling the lumen (Figure 2.11). There are more outer tubules in the knob than in the tentacle shaft. This arrangement of two concentric rings of microtubules is common to most suctorian tentacles, although the number of microtubules involved varies from species to species. In *Heliophrya erhardi*, for example, the number of microtubules in the inner ring exceeds 200, which is probably near the maximum (Spoon *et al*., 1976). The inner row of tubules possess arms which project from the luminal surface as shown in Figure 2.12, in feeding and non-feeding tentacles. When a prey contacts a tentacle, toxicysts characteristic of suctorians called haptocysts are discharged releasing enzymes which immobilise the prey. Prey may perform defensive behaviour. *Paramecium*, for example, discharges trichocysts at the point of contact, which has the effect of uprooting the cilia attached to the knob. Firmly attached prey can sometimes escape by turning rapidly and twisting off the tentacle or a portion of their own isolated cytoplasm (Spoon *et al*., 1976).

Various theories have been put forward to explain how food is transported along the tentacle. Early theories proposed that the motive force was generated within the body of the suctorian by increased contractile vacuole activity, or by reduced hydrostatic pressure relative to the prey. Alternatively peristaltic waves of contraction along the tentacles and retraction of cytoplasm from the lumen of the tube have also been suggested. We now know that the mechanism resides in the tentacle itself and is mediated by the microtubular elements (Bardele, 1972; Tucker, 1974).

The process of ingestion is preceded by changes in the structure of the tentacle, as shown in Figure 2.12. Prior to ingestion the unit membranes of the prey and predator are closely applied. As feeding commences the epiplastic rim increases its diameter and moves downwards, as shown in Figure 2.12b. The microtubules in the terminal knob bend along their longitudinal axes and splay apart at levels above the sleeve as their tips move downwards and outwards. Thus a large area of knob is extended into the prey, presumably facilitating the uptake of material.

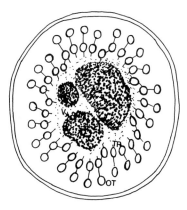

Figure 2.11: A cross-section of the Shaft of a Resting Tentacle of *Tokophrya*. OT — outer tubules, TR — inner tubule ring, V — dense vesicles in the lumen of the tentacle.
Source: Based on Tucker (1974).

Figure 2.12: The Terminal Portion of the Tentacle of the Suctorian *Tokophrya*. The non-feeding tentacle is on the left and the feeding structure of the tentacle is shown on the right. cm — cell membrane, el — endoplasmic layer, mv — membranous invagination, ot — outer tubules, pm — prey cell membrane, sm — suctorian cell membrane, tr — inner tubules.

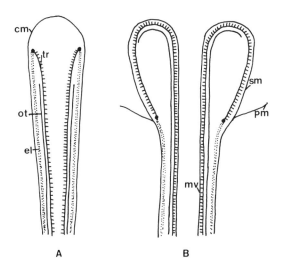

A   B

Source: Based on Tucker (1974).

The surface membrane invaginates down into the tentacle forming a tube for the passage of material (Tucker, 1974). Tucker (1974) suggests that the invagination moves downwards throughout ingestion, taking prey cytoplasm and organelles with it, and this is brought about by actively contractile elements in the tentacle. The prey contents are thus transported into the predator where digestive vacuoles are formed. The process is not one of suction, but is more aptly described as a grasp-and-swallow mechanism (Bardele, 1972). It is not certain how the invaginated membrane is moved downwards, but the evidence suggests that the arm-bearing microtubules are involved in moving the cytoplasmic invagination downwards (Bardele, 1974; Tucker, 1974). The invaginating plasma membrane must be used up internally in the formation of digestive vacuoles and must be replaced. Vesicles moving up the tentacles extraluminally have been implicated in performing this role, their function being analogous to the discoidal vesicles described earlier in digestive vacuole membrane recycling in other ciliates (Rudzinska, 1970; Tucker, 1974).

(ii) Protozoa Lacking a Cytostome

Among the free-living Protozoa lacking a cell mouth are the Sarcodina and the majority of the flagellates. Each of these groups has evolved a variety of methods for ingesting their food. Their diet includes bacteria, algae, Protozoa and some elements of the micro- and meiofauna.

Naked amoebae, for example *Amoeba proteus* and *Chaos carolinense*, engulf food material large or small, by flowing around it and enclosing the food in a vacuole with some of the external liquid medium. A pseudopod is locally checked when it touches a particle and thus flows around it. Rhumbler (1910, quoted in Kudo, 1971) observed four methods of food ingestion in amoebae. In what is described as 'circumfluence', the cytoplasm flows around the food organism as soon as it makes contact with it, while in 'circumvallation' the amoebae form pseudopods around a food vacuole without first touching it. No contact between amoebae and food occurs before ingestion in 'circumvallation', the stimulation in this case being mediated chemically. Ingestion may result by 'import', in which the food moves into the amoeba with little movement taking place on the part of the protozoan, or 'invagination' of food may occur when an amoeba touches and adheres to the food, causing the ectoplasm to contract and invaginate as a tube which becomes pinched off as a food vacuole. Many of the larger amoebae, for example *Amoeba proteus*, are carnivorous on other Protozoa, while small amoebae exploit bacteria and algae, and possibly also detrital material.

The feeding biology of amoebae bearing internal and external skeletal structures is extremely varied, and some very specialised feeding mechanisms have evolved. Radiolarian sarcodines take a wide variety of food, including flagellates and small crustaceans such as the brine shrimp *Artemia* and various species of copepods. When a flagellate comes into contact with the halo of axopodia radiating from a radiolarian, the prey is quickly immobilised. Adherence to the axopodium is mediated by a mucus-like substance secreted by Golgi-derived vesicles. Following adhesion and immobilisation, the prey is transported down the axopodium into the main body of the cell. Crustacean prey present more of a problem to radiolarians because they are large, have mobile strength and the nutritive material is enclosed in an exoskeleton. When contact between an axopodium and a crustacean occurs, the prey is immediately entangled. As the crustacean struggles it becomes more entangled in the strands of the axopodia, which orientate their streaming so as to maximise the number in contact with the prey. The axopodia flow along the broad surfaces of the exoskeleton and engulf the appendages. The forces exerted by the engulfing axopodia are such that eventually the exoskeleton ruptures, and the axopodia are able to penetrate the soft tissues of the crustacean and pry pieces off which are then directed down the axopodia to the main body of the cell (Anderson, 1980).

Heliozoans feed mainly on flagellates, ciliates and small metazoans which have the misfortune to swim onto the long radiating axopodia of these protozoans. The process of food capture and ingestion has been described in *Actinophrys sol* feeding on the ciliate *Colpidium campylum* (Patterson and Hausmann, 1981). The random contact of a prey with an axopodium results in capture by adhesion of the cilia to the heliozoan. The adhesive substance appears to be on the heliozoan and it is suggested that extrusomes associated with the plasma membrane are involved. Extrusomes may also play a role in digestion and in the production of digestive vacuole membrane in these amoebae. The prey is drawn closer by arm resorption in the region of capture. Within five minutes of capture the heliozoan produces fine active pseudopodia from the region of the cell body adjacent to the prey. These develop so that a funnel-shaped pseudopodium is produced which is pulled over the prey like a sheath by the action of the leading margin. As the pseudopod passes over the ciliate, the opening constricts until the prey is completely enclosed in a food vacuole. The whole process requires about 20 minutes (Figure 2.13). Frequently other individuals of *Actinophrys* may fuse with the feeding individual, separating again on completion of digestion. This represents a useful adaptation allowing cooperation in the feeding on, and processing of, a prey too large for a single heliozoan to cope with.

Figure 2.13: A Heliozoan Ingesting a Ciliate Prey.

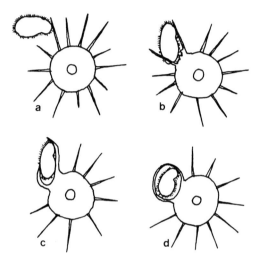

Source: Based on Patterson and Hausmann (1981).

Foraminiferans also exploit a wide range of food. Some, for example *Mansipella arenaria*, are suspension feeders taking small particles including bacteria, algae and fine detritus. Others are benthic, engulfing bacteria and detritus from the sediment. Some exotic modes of feeding have evolved among carnivorous species which prey on micro- and meiofauna. *Pilulina argentea* builds a deep bowl-shaped test with a large opening on the surface of the mud. Essentially *Pilulina* is a living pitfall trap. The surface of the test opening is covered with a 'roof' of sticky pseudopodia camouflaged by a fine layer of adhering mud. Any small copepod that blunders onto the 'roof' gets stuck and is drawn into the animal. *Bathysiphon*, an erect benthic foraminiferan, has knobs of sticky protoplasm emerging from the test surface. The sticky pseudopodial knobs capture prey and process the prey in similar fashion to suctorian ciliates (Lee, J.J., 1980). Testate amoebae feed on bacteria, fungi and unicellular algae, engulfing particles in typical amoeboid fashion. However, there are testate amoebae, for example *Pontigulasia* and *Lesquereusia*, which feed on filamentous algae like *Spirogyra* by piercing each algal cell and extracting the cell contents. The algal cell wall is broken down either enzymatically or by pulling with attached pseudopodia. Following this the pseudopodia are extended into the algal cell where they tear off and ingest portions of the cell contents (Stump, 1935).

Although many of the heterotrophic colourless flagellates possess a cytostome (see Section B(i)), there are members of the Mastigophora feeding on particulate material which lack a mouth. Among these are the collared flagellates, so called because the flagellum is surrounded by a delicate collar. Particles are driven against the collar by a flagellum-induced water current. Food particles collect on the collar which then contracts or partially rolls up, bringing the collected material into contact with the cell surface, where food vacuoles are formed below the base of the collar. The collar is constructed rather like a net made up of numerous finger-like processes which are connected to each other by micropseudopodia (Fjerdingstad, 1961). Other acytostomate flagellates, e.g. *Monas* sp., form food vacuoles by pseudopodial action near the base of the flagellum. The flagellum creates a water current which directs food onto the flagellar base area.

(iii) Pinocytosis

Many free-living Protozoa are capable not only of ingesting particles, but also of taking in fluid droplets of the medium as vacuoles or vesicles in a process known as pinocytosis or 'cell drinking'. Mast and Doyle (1934) described the phenomenon in *Amoeba proteus*, *A dubia* and *A. dolfeini*. When these species were placed in weak egg albumen solutions, locomotion ceased, the contractile vacuole stopped pulsating and the surface of the cell began to wrinkle. This latter process continued until the cells were covered by protuberances, crevices and folds. Channels a few micrometres wide and between 15-50 micrometres long developed in the cytoplasm from which vesicles fragmented into the cell. These authors suggested that pinocytotic activity in amoebae serves to compensate for a loss of water to the medium. Later work by Chapman-Andersen and Prescott (1956) on *Chaos chaos* and *Amoeba proteus* confirmed this pattern, though they suggested that pinocytosis may serve an additional nutritional role. Pinocytosis is performed by *Tetrahymena* when cultured axenically in proteose-peptone medium. Food vacuoles are formed in the normal manner by the cytostome and are composed of droplets of nutrient solution.

Among free-living Protozoa pinocytosis may be an aberrant activity since it appears to be induced only in solutions which differ radically from the normal aquatic environment, and it is probable that this form of nutrient intake is not a characteristic mode of nutrition in the wild. There is no doubt that in many parasitic species pinocytosis is the normal mode of feeding. Experiments on pinocytosis in free-living Protozoa indicate that the ability to pinocytose is present though not necessarily functional under normal circumstances, but many species which have evolved a parasitic way of life have come to rely on pinocytosis for nutrition.

C. Food Selection in Protozoa

The concept of food selection being practised by single-celled animal-like organisms, which lack an apparent sensory system, may seem somewhat far-fetched. There are, however, numerous examples of food selection practised by protozoans, although in most cases the exact mechanism controlling how a protozoan selects one item of food in preference to another is not yet understood in detail. Bacterivore filter-feeding ciliates practise no selection as far as the palatability of particles is concerned; it is the size of the particle alone which determines whether or not it is ingested. Each species has its own size range of particle and this is determined by the morphology of the mouth, mainly by the size of the free spaces between the cilia on the organelles which act as the sieve (Fenchel, 1980a, 1980b). In some species the sieve is the adoral membrane, in others the membranelles (see Section B(i)). Among bacterivores there appears to be a varying growth response to various bacterial species. Curds and Vandyke (1966) cultured five species of ciliate monoxenically, that is in culture with one species of bacterium and one species of protozoan, on 19 strains (15 species) of bacteria. Some of the bacterial strains were toxic, others although not toxic did not support growth, while some bacteria allowed good ciliate growth and reproduction. Thus it would appear that in some respects the feeding of such filter-feeding bacterivores is rather haphazard because a proportion of the ingested material may be of little or no energy value. The fact that various species exploit different size spectra of food particles explains in part why so many ciliates are able to cohabit in protozoan communities, since they will not necessarily be exploiting the same bacteria as food.

Food selection in *Stentor* was investigated during the early part of this century by Schaeffer (1910). From his elegant experiments Schaeffer showed that this large omnivorous ciliate exercises selection among the particles that are brought to its cytostome by the ciliary current. The selection is mediated by changes in the beat of the cilia of the oral area. Of any group of particles arriving in the oral region in the feeding current some will be carried to the cytostome and ingested, while others will be rejected by localised reversal of ciliary beat. *Stentor* is able to discriminate between different food organisms, eating some and rejecting others. Moreover, it can distinguish between animate and inanimate particles such as glass, carmine and starch. The degree of selectivity in feeding appears to vary depending on the nutritional state of *Stentor*. The ciliate discriminates more perfectly when almost satiated than when hungry. Later investigations have shown a preference for ciliate prey over algal or flagellate prey (Hetherington, 1932; Rapport, Berger and Reid, 1972).

Exclusively carnivorous ciliates exercise very distinct selection during feeding. Suctorians feed almost entirely on holotrichous and spirotrichous ciliates, except those holotrichs with a hard pellicle, for example *Coleps*. The ciliated dispersal young or swarmers of suctorians are never captured, nor are hypotrich ciliates, flagellates or amoebae. Hull (1954, 1961a) demonstrated that agar models of prey coated with fresh ciliate homogenate were captured as were various model prey coated with acetylcholine. The evidence points to a chemical reaction between compounds on the prey surface and the tentacle tip. Presumably those protozoans which are not retained when they make contact with suctorian tentacles, lack the appropriate chemical make-up on their cell surface.

The ciliate predator *Didinium*, which is generally believed to feed exclusively on species of the genus *Paramecium*, can be made to prey on other ciliates experimentally. Clones of didinia which feed on *Colpidium* have been produced (Berger, 1979). The *Didinia* become 'imprinted' on this smaller prey species, although they still retain a diminished ability to attack *Paramecium aurelia* and to a lesser extent *P. caudatum*. However, they reproduce poorly and the clone quickly dies out on a diet of paramecia. Other reports substantiate the ability of *Didinium* to survive on non-paramecid prey (Seravin and Orlovakaja, 1977). The indications are that the imprinting and recognition of prey is chemical. Whether *Didinium* resorts to other prey in the wild is a matter of some conjecture. Normally when their paramecid prey are exhausted they encyst until food again becomes available, whereupon they excyst. Laboratory studies of this type, however, are valuable in elucidating the mechanism by which Protozoa perform selection of food. Other predaceous protozoans, including *Dileptus*, *Amoeba proteus* and *Peranema* use the chemical characteristics of the prey for recognition of suitable food items. Certain potential prey are capable of defence by the release of trichocysts, although this behaviour is not always effective in preventing capture. Seravin and Orlovaskaja (1977) propose that whether a protozoan is captured or not may depend on an interaction of attractants and repellants given off into the environment, although as yet the evidence is insufficient to substantiate this hypothesis.

Sarcodina exhibit food selection, although the phenomenon is less well documented in these protozoans. *Amoeba proteus* will select *Chilomonas paramecium* in preference to *Monas punctum*, even when there are many more *Monas* available in the medium. *Monas* appears to be less easily digested, taking up to 3½ hours to die in food vacuoles compared with 3-18 minutes for *Chilomonas* (Mast and Hahnert, 1935). Anderson (1980) has described food selection in radiolarians. Two methods of rejecting unsuitable particles have been observed. Some algae are rejected immediately on contact, with no adhesion to the axopodia taking place. In the other case the prey is apprehended and enclosed in a vacuole, but after several minutes the immobilised cells are released, often in large quantities simultaneously. This delayed response suggests that some form of physico-chemical analysis is performed by the sarcodine during the temporary engulfment of food material.

D. Digestive Processes

When food particles enter a protozoan cell either by phagocytosis at the membrane or via a cytostome, they are enclosed in a membrane with some of the external medium as a food vacuole. The digestion of food in vacuoles is not unique to protozoans – it is also the sole means of digestion in Porifera and occurs to some extent in Cnidaria, Ctenophora, Turbellaria, Rotifera and Brachiopoda. Essentially the food vacuole of a protozoan is analogous to the gut of higher organisms, but instead of the permanent tissue structure of an alimentary tract, the membrane materials of the digestive vacuoles are recycled in the cell from cytoproct to cytopharynx.

Typically, digestive vacuoles decrease in diameter and the enclosed particles become aggregated as digestion proceeds. During the digestive cycle in *Paramecium* the vacuoles decrease in size by about 50 per cent during the first seven minutes, the reduction in size being attributable to the removal of a large percentage of the membrane by endocytic-like processes. Subsequently the vacuole expands as the result of the fusion of the digestive vacuole with lysosomal membrane (Allen and Staehelin, 1981) – Figure 2.14. The initial shrinkage of the vacuole may be the result of the recovery of specific membrane components to meet the demand for vacuole membrane formation at the cytostome. This does not contradict the membrane recycling process which has been observed in *Paramecium* (Allen, 1974; Allen and Fok, 1980). Such material may complement the membrane material provided by the normal recycling of membrane. The digestion of food in the vacuole does not commence until the vacuole begins to expand. In other Protozoa there may not be a reduction in vacuole size followed by an increase. Instead, as in the heliozoan *Actinophrys*, the contents of the vacuole coagulate during the first hour, and in the following two hours the fluid in the digestive vacuole is resorbed. During the vacuolar condensation phase in *Actinophrys* numerous small vesicles accumulate around the periphery of the vacuole, presumably playing a role in digestion (Patterson and Hausmann, 1981). Similarly in *Climacostomium virens*, an algivorous ciliate, the vacuole condenses, after lysosomes surrounding the vacuole fuse with it (Fischer-Defoy and Hausmann, 1982).

During digestion the contents of the digestive vacuole become progressively more acid as enzymatic action proceeds, but eventually change to alkaline. In *Paramecium caudatum* there is a rapid reduction in pH as soon as vacuole release occurs, dropping from pH 7 to around pH 3 in five minutes, then increasing again in the neutralisation process to the original levels within 11 minutes of vacuole formation. The mechanism of the acidification process, however, is as yet unknown (Fok *et al.*, 1982). In *Paramecium caudatum* the digestive cycle is fairly short; after 40 minutes about 90 per cent of the cells have completed digestion and defaecated. Obviously the length of the digestive cycle is a function of the type of diet exploited and probably also prevalent temperatures, although the latter aspect requires elucidation. In the algivorous species *Climacostomum* the whole process from vacuole formation to egestion of waste material may take up to 24 hours (Fischer-Defoy and Hausmann, 1982) and in the carnivore *Actinophrys* twelve hours are required to digest the ciliate *Colpidium* (Patterson and Hausmann, 1981).

Hydrolytic enzymes functioning in the digestive processes of the cell have been reported in protozoans. Among these are nucleases and acid

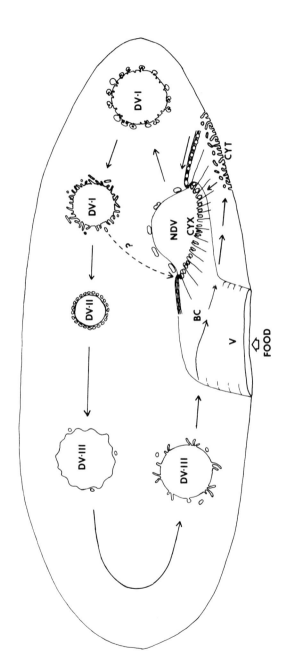

Figure 2.14: The Phases of Digestion in *Paramecium*. BC — buccal cavity, CYT — cytoproct, CYX — cytopharynx, DV — digestive vacuole, NDV — nascent digestive vacuole.

Source: Allen and Staehelin (1981). Reproduced from the *Journal of Cell Biology*, 1981, *89*, by copyright permission of The Rockerfeller University Press.

phosphatases which split high molecular weight nucleic acids, proteinases and peptidases responsible for splitting proteins into amino acids, esterases and lipases which act on fatty acids and carbohydrases functioning on polysaccharides. Although there is still much research to be done in this area of protozoan biochemistry a body of information is accumulating to substantiate the specific activities of hydrolases in protozoan cell and food vacuole digestive processes.

The evidence suggests that the appearance of enzymatic activity in a food vacuole is not dependent on the synthesis of new enzymes. It appears that enzymes are present in appreciable quantities prior to vacuole formation during feeding, and are transported to the vacuoles, because acid phosphatase levels in some species are the same before and after the formation of food vacuoles. After vacuoles are formed there is an accumulation of small granules showing acid phosphatase activity in the cytoplasm around each vacuole. Such hydrolytic enzymes have been demonstrated in a large number of species. The site of enzyme production is probably the ergastoplasm, which is well developed in most Protozoa. The Golgi apparatus, if present, may participate in the condensation of enzymes (Muller, 1967). Hydrolytic enzymes appear to fall into two functional groups in Protozoa, those which are engaged in digestive processes occurring within the cell and others which function in the vacuoles.

Muller (1967) has speculated on the evolution of digestive processes in Protozoa. The widely-accepted view is that intracellular digestion is the ancient form of food breakdown and the extracellular cavity digestion of food as seen in metozoans evolved from intracellular digestion only after the formation of multicellular organisms. This view fails to explain several aspects of protozoan digestion, including the possession of extracellular hydrolases. An opposing hypothesis generated by Ugolev (quoted in Muller, 1967) is based on the fact that all types of organisms, both multicellular and unicellular, are able to release enzymes into the medium. This ability may have had no functional value, originally being the result of leakage through imperfect membranes; it declined during evolution but did not entirely disappear. When endocytic processes produced vacuoles within the cell, hydrolases leaking through the membranes of such vacuoles found a favourable environment for activity. The evolution of this process led to the development of a lysosomal system involving the appearance of specific enzyme carriers and a specialisation of enzymes, thus leading to the preservation and intensification of enzyme secretion through cell membranes in the digestive tract of Metazoa. Two types of enzyme necessarily develop in such a system; those participating in extracellular digestion and those involved in intracellular digestion. This hypothesis argues that both types of digestion evolved simultaneously and progressed in different directions. Muller maintains that this view explains the presence of both types of digestion in Protozoa and accounts for the differences between intracellular and extracellular digestion.

Among higher organisms there are two broad categories of digestion and defaecation strategy employed. Some organisms carry out continuous flow digestion through what can be described as a single compartment gut in which there is no mixing of the food, no selective retention of less digestible particles and the ingestion rate is equal to the egestion rate. Other animals carry out discontinuous flow through a one-compartment digestive system. In this case the rates of ingestion and defaecation are not equal, so that animals that feed by day, for example, show an exponential decrease in the defaecation rate after they cease feeding. It follows that in continuous flow digestion the rate of flow of material through the alimentary tract must be constant throughout the system, whereas in discontinuous flow digestion the throughput rate is variable (Sibly, 1981). A complication arises in the case of animals having continuous flow digestion through a gut with additional compartments for cellulose fermentation.

Do Protozoa fit into a pattern of digestion and defaecation as do higher heterotrophic organisms? The question is complicated by the problem of defining what is the equivalent of the alimentary tract in protozoans. Traditionally the food or digestive vacuole has been regarded as analogous to the gut of higher organisms. Recent ultrastructural work has shown, however, that some species of ciliate possess a pool of continuously recycling membrane material which forms the digestive vacuoles (Allen, 1974; Allen and Fok, 1980; Fischer-Defoy and Hausmann, 1982). In such protozoans one could argue that the pool of membrane material is analogous to the alimentary tract of metazoans and each digestive vacuole contains a discrete meal passing through the gut. Obviously there is no mixing of 'meals' in Protozoa, although there is evidence that food vacuoles 'exist' for variable lengths of time in an individual, depending on the digestibility of their contents, so that there may be a disparity between the rates of ingestion and egestion in species exploiting a mixture of food organisms. Protozoa exploiting a uniform food source, such as a single bacteria species, probably have similar rates of ingestion and egestion and are therefore comparable to metazoans with a continuous flow digestive system. Carnivorous protozoans such as *Didinium nasutum* and *Actinophrys sol* are discontinuous feeders in that they take one large prey at a time and process that meal before taking another. In this case one ingestive effort leads to one defecation, so that the rates of ingestion and egestion are equal; which does not conform to higher predators.

Some multicellular heterotrophs show variation in gut length, and therefore volume, as a function of seasonal variations in the quality of available food. Birds, for example, extend the gut during periods when the diet has a high indigestible plant component (Sibly, 1981). The ability to modify the area of functioning tissue or cellular material for the digestion of ingested energy is a characteristic the Protozoa share with some higher organisms. In the Protozoa, however, the variation is short-term and a function of food availability and the rate of ingestion. Multicellular animals have to balance the advantages and disadvantages of increasing gut size; a larger alimentary tract allows greater digestive efficiency, but it has to be carried around with its contents, which imposes a weight problem and reduces speed of movement leading to greater susceptibility to predation. Protozoa do not have to contend with this problem. In general, herbivorous animals have longer guts than carnivores and a parallel can be found in the Protozoa. Carnivores such as *Didinium nasutum* or *Actinophrys sol* take one large prey, enclosing it in a single vacuole. The area of membrane material formed into digestive vacuoles in such carnivores will be considerably less than the amount involved in the numerous food vacuoles found in bacterivores and those algivorous species which exploit unicellular algae.

At present our knowledge of the feeding behaviour of protozoans, particularly feeding periodicity, and digestive processes in relation to the digestibility, energy content and availability of food, is limited. Future research may allow us to define strategies of feeding and digestion as a function of food supply and environmental conditions in these unicellular creatures, which will in turn give us greater insight into one aspect of how the physiology of Protozoa relates to their ecology.

E. Symbiotic Relationships in Protozoa

Symbiotic relationships exist between free-living Protozoa and some fungi, but more commonly with bacteria and various algae. Among the latter are some blue-green algae, referred to as cyanellae, green algae termed zoochlorellae, and dinoflagellates usually called zooxanthellae. A large number of ciliate and sarcodine species carry such symbionts. They are less common in flagellate Protozoa. A range of protozoan

CAMROSE LUTHERAN COLLEGE
LIBRARY

species involved in symbiotic associations with algae has been reviewed by Curds (1977). Undoubtedly the algae contribute some of the products of photosynthesis to their hosts and the protozoan probably provides some chemical material in return, in addition to shelter. The exact chemical basis and the extent of these relationships are still rather unclear. Bacteria as endosymbionts are less well understood, although there are numerous reports of bacteria or suspected bacterial symbionts in Protozoa (Grell, 1973; Curds, 1977). Recently endo- and ectosymbiotic bacteria have been reported in a unique group of anaerobic ciliates, the 'sulphide ciliates'. The bacteria are believed to utilise the metabolic products of the ciliate metabolism (Fenchel et al., 1977).

F. Factors Influencing Feeding

The amount of energy ingested and the rate of ingestion are influenced by both biological and environmental factors. Among the biological factors is the type or species of food available. Protozoa exercise selectivity in their feeding behaviour; thus when the preferred species in the repertoire of food exploited is sparse or absent there will naturally be an impact on the rate of energy ingestion and the overall quantity consumed. Where several protozoan species have overlapping food preferences, as will certainly be the case in many communities, the dimension of competition enters as a factor playing a role in determining feeding rate. The concentration of food and the density of the protozoans grazing or preying on that food may have a profound effect on the rate of food intake.

Temperature is one of the most important environmental factors influencing the physiological activities of ectothermic organisms. The feeding activity and metabolic rate of Protozoa are often a reflection of prevalent ambient temperatures. As temperature falls the energy demands which must be met by feeding decrease. Other environmental factors such as pH and oxygen availability probably also affect the energetics of feeding, but unfortunately the impact of these characteristics is poorly documented.

Different species show variable responses to fluctuations in the concentration of their food supply. Feeding rate in bacterivore ciliates appears to be a function of food concentration at lower bacterial densities, but becomes independent of the concentration at high bacterial cell density. Harding (1937) was one of the first researchers to consider feeding rate and bacterial food supply in relation to growth and reproduction. His studies on the ciliate *Glaucoma* feeding on *Pseudomonas* showed a clear relationship between feeding rate and food supply at low bacterial concentrations. The rate of ingestion reached a maximum level as bacterial density was increased, beyond which high food concentrations produced no response in feeding rate. *Colpidium campylum* grazing on *Moraxella* sp. showed the same pattern of response to bacterial density (Figure 2.15; Laybourn and Stewart, 1975). The ingestion rate of *Colpidium* appears to be influenced solely by food availability, because temperature had no significant impact between 10°C and 20°C. Clearly there is a maximum rate at which these ciliates can feed and exploit the energy available to them. Many of the higher bacterial densities used in experiments designed to measure ingestion are probably in excess of those encountered in the field, where in any case there is usually a heterogeneous bacterial flora, some of which will be unpalatable to any given protozoan species. It is likely that the densities of available food in the natural environment rarely allow the maximum feeding rate to be achieved.

Figure 2.15: Bacterial Consumption by *Colpidium campylum* at Different Food Densities at 10°C (●), 15°C (○) and 20°C (▲).

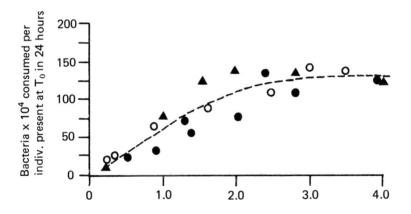

Source: Laybourn and Stewart (1975), with the permission of the British Ecological Society.

Figure 2.16: Predation on *Tetrahymena pyriformis* by *Amoeba proteus* in Relation to Temperature and Prey Density. ● -10°C, ▲ -15°C, ■ -20°C.

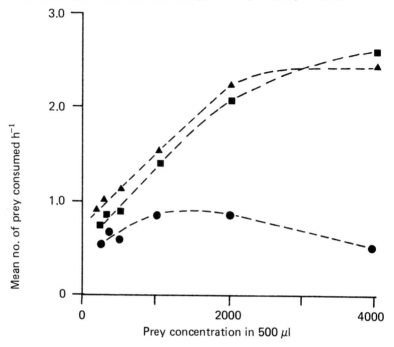

Source: Data from Rogerson (1981).

Sarcodinids exploiting bacteria and other small cells, such as yeasts, present a slightly confusing picture. *Acanthamoeba* sp., grazing on cells of the yeast *Saccharomyces cerevisiae*, had a feeding rate linearly related to food supply below a ratio of one amoeba to 200 yeast cells; above this food concentration, however, the rate of ingestion was very variable and showed no obvious relationship with the density of yeast (Heal, 1967). The higher experimental densities of both food and amoebae were probably atypical of the soil communities in which this particular species of *Acanthamoeba* normally lives, although the lower yeast densities may be a more realistic reflection of the natural situation. Predaceous amoebae also show a more or less linear relationship between ingestion rate and prey density. *Amoeba proteus* preying on *Tetrahymena* in densities of 125 to 4,000 prey per 500 μl drop of medium was able to exploit progressively increasing prey populations up to a maximum rate which was sustained over higher densities but then declined at very high prey concentrations (Figure 2.16; Rogerson,

1981). In *Amoeba proteus* the feeding rate is strongly influenced by temperature, as Figure 2.16 shows. Maximum feeding rate at 10°C occurs at a density of 1,000-2,000 *Tetrahymena* per amoeba and at 15°C at a density of 2,000-3,000. At 20°C the prey concentration was extended to 8,000 to each amoeba. The effect of this very high prey density is to depress capture rate dramatically, to less than one-quarter of the maximum rate. The value of experimental data at high prey densities in a confined space is questionable. A decline in the feeding rate of *Amoeba proteus* in dense prey populations suggests physiological stress, induced by the density of the material surrounding the amoeba imposing a mechanical effect, together with the chemical impact of the accumulation of metabolites in the medium. Functional value in relation to the wild populations can only be placed on data derived from lower ranges of prey density.

Benthic foraminiferan species belonging to the genera *Allogromia*, *Quinqueloculina*, *Elphidium*, *Rosalina* and *Globigerina* feed on algae and bacteria. Studies of their feeding behaviour have indicated that a combination of species and age of food organism, age of foraminiferan and food concentration all influence the feeding rate. Small individuals of *Allogromia* between 150-200 μm in diameter, for example, consumed 200 per cent more food than their larger older relatives which were 350-400 μm in diameter (Lee *et al.*, 1966). Possibly the energy demands of older slower-growing individuals are lower and this is reflected in the feeding rate. A maximum of 10^4-10^5 organisms were eaten by the foraminiferan species in the experiments carried out by Lee *et al.* (1966) and the feeding rate was a linear function of food density.

Among the carnivorous ciliated Protozoa, *Didinium* is perhaps one of the most voracious and spectacular feeders. Salt (1974) found that the rate of prey capture was related to prey density in experiments where *Didinium* was offered an increasing number of prey. When the density of didinia was increased relative to prey there was a decrease in the capture rate as more predators competed for the same number of prey. Thus both predator and prey densities play a role in determining feeding rate. The question arises as to whether the control of capture rate by prey and predator density is fixed or variable. Salt believes that they are alternatives and the capture rate is set by whichever is the more restrictive. For example, let us assume a predator–prey combination of two *Didinia* and 50 *Paramecia*. The predator density control would set the capture rate at around 1.7 prey per *Didinium* per hour, while the prey density control would set the capture rate at 1.4

paramecia per *Didinium* per hour. As the density of *Didinia* increases, the rate set by the prey density control decreases, but if the density of prey were restored to 50 as the predator density passed six then the regulation would pass to the predator control. The evidence suggests that the prey density control is fixed, but it is not yet possible to determine if the predator density is a fixed or variable control. Contradictory evidence exists on the relationship between prey density and capture rate in some suctorians. Rudzinska (1951) found no limit to the amount of food consumed in *Tokophrya infusionum*. The number of progeny produced was a function of the quantity of food eaten. One prey item resulted in the production of one embryo and two to three prey in two embryos. At high prey densities, however, the number of offspring produced declines so that where 40 prey are consumed in 24 hours only one young results. Interestingly, *Tokophrya* which fed continuously in Rudzinska's experiments turned into giants. It is unlikely that suctorians feed continuously in their natural habitats except in super-abundant prey densities which are probably rare. The dependence of these sessile protozoans on chance collisions between their tentacles and swimming prey implies that the opportunity to feed is limited. Another suctorian, *Podophrya fixa*, preying on varying densities of *Colpidium campylum*, showed reduced rates of capture at prey densities exceeding 30 colpidia per 50 μl. Observations suggest that *Podophrya* becomes satiated after three to four prey have been fed upon. A larva is then produced and feeding recommences as more prey are captured (Laybourn, 1976a).

The influence of temperature on the feeding rate of various groups of Protozoa varies considerably. In *Stentor coeruleus* the feeding rate on *Tetrahymena* was the same at 15°C and 20°C (Laybourn, 1976b) and in *Colpidium campylum* temperatures between 10°C and 20°C had no significant impact on the role of bacterial ingestion, as Figure 2.15 shows (Laybourn and Stewart, 1975). The testate amoeba *Arcella vulgaris* also consumed bacteria at the same rate irrespective of temperature within the range 10-30°C (Laybourn and Whymant, 1980). While these species appear to have a feeding behaviour unaffected by temperatures, there are reports of ciliates and amoebae which have their feeding rates modified by environmental temperature variation. In *Paramecium*, for example, the production of food vacuoles increased two-fold between 10°C and 20°C (Metalnikow, 1912). The number of bacteria enclosed in each food vacuole was not recorded, so that the actual quantity of food ingested may or may not have varied. A similar study using food vacuole formation as a criterion of feeding rate on the folliculinid *Parafolliculina amphora* at temperatures of 16°C-28°C, showed a doubling in the rate of vacuoles formed with an increase of 10°C in accordance with van't Hoff's Law (Andrews, 1947). The overall influence of temperature on feeding rates is difficult to gauge with any degree of accuracy, because of the variables involved and the different criteria used to estimate ingestion rates. There are numerous modes of feeding in protozoans so that one particular method of feeding may be temperature influenced where another may not. While *Amoeba proteus*, for example, has its ability to capture prey modified by temperature (Rogerson, 1981), the ciliate *Stentor* feeding on the same prey species does not (Laybourn, 1976b). *Amoeba proteus* and *Stentor* use entirely different methods of ingestion.

Clearly feeding rate is influenced by a combination of factors acting together, so that in some species it is difficult to implicate any one isolated factor as responsible for variation in the rate at which energy is consumed. One of the major problems is the impossibility of conducting complex investigations on feeding rate in the field. While it is feasible to undertake limited studies as Goulder (1973), for example, did on the possibility of diurnal variations in feeding on algae by *Loxodes magnus* and *L. striatus*, large-scale investigations are logistically difficult. A dilemma exists in attempts to apply laboratory data to field communities, and until more detailed information on the structure of decomposer communities, especially in aquatic environments, and competitive interactions between the elements of protozoan communities, becomes available, no real clear picture of regulatory mechanisms in feeding rates will be achieved. For the moment we must content ourselves with the limited information we can derive from laboratory investigations and accept that they are probably a limited reflection of the natural situation.

3 PHYSIOLOGICAL FUNCTIONING OF PROTOZOA

A. Introduction

The investigation and characterisation of protozoan physiology often requires an approach which differs markedly from that applied in metazoan studies. Quantifying physiological function in small organisms requires refined micro-techniques, and over and above this technical problem, there are a number of complications which occur in protozoological studies which do not pertain in physiological investigations on most metazoans. First, most protozoans have relatively short life-cycles, particularly at higher temperatures, so that growth and division arc rapid, and there is the added complication of variation in the cell size attained before the initiation of division in each species. Secondly the food organisms of the majority of Protozoa also undergo rapid division, which can be an important factor where food concentration is critical in a physiological investigation.

The usual asexual life-cycle of protozoans differs in a fundamental manner from that of metazoans, because in protozoans growth and reproduction are inseparable. In metazoan animals there is usually a developmental period during which growth is high, followed by an adult phase where growth is minimal or non-existent, and all energy for production is directed into reproduction. Having reproduced once, or repeatedly, the adult eventually dies and the corpse is lost to the biomass of that particular species population. By contrast, in the Protozoa, growth and reproduction are inseparable in energetic terms, because during reproduction, the whole body or cell is the reproductive product. The cell mass is passed on as two new individuals to the succeeding generation and the biomass is retained within the population. In theory the biomass of a given population could go on increasing in an exponential fashion indefinitely. In reality, of course, environmental factors and biological factors mediate against such an event.

A wide diversity of physiological function is observable in the Protozoa, and is in part the consequence of the polyphyletic origin of the phylum and the various stages of evolutionary specialisation which

66

have been achieved within the protozoan groups, but it is also the result of the diverse habitats colonised by protozoans. Each type of environment offers a different spectrum of environmental factors such as temperature regime, pH, oxygen availability, salinity and sediment type, which will induce variations in physiological performance, even among closely-related species.

B. Asexual Life-cycle

In free-living Protozoa the typical life-cycle involves growth or increase in cell size followed by binary fission or some other form of asexual reproduction. As a general rule free-living species resort to sexual reproduction in adversity, when the environment becomes unfavourable in some way or if the food supply declines. The exact factors responsible are unclear, and probably vary from one species to another depending on their ecological tolerances. An almost exclusive reliance on asexual reproduction is common in the Sarcodina and to a large extent in the Mastigophora, but rare in the ciliates where almost all species possess the ability to reproduce sexually if necessary. This discontinuity in the distribution of asexuality is usually explained by the hypothesis that sexuality has evolved in some groups, but not in others. However, it has been suggested that total asexuality in some protozoan groups is the result of the loss of sexual ability and that this phenomenon has occurred independently in various classes and orders of Protozoa (Hawes, 1963).

During the growth and division cycle of a protozoan there must necessarily be a phase of DNA synthesis and replication of chromosomes, as well as a phase of cell growth. The production of DNA usually occurs at a particular phase in the cell life-cycle, and normally at a different time in the cycle from the division of the nucleus. The phases of the cell cycle have been described by Prescott and Stone (1967). Four phases are discernible in the normal cell cycle: First a division phase, secondly a phase which spans the period from the end of the division to the beginning of DNA synthesis which is essentially a growth phase, thirdly the period of DNA synthesis, and lastly a phase which extends from the end of DNA synthesis until the beginning of the next division. The timing of these phases in the life-cycle and their length varies depending on species and on the availability of essential amino acids in the normal diet or artificial medium.

The division of a protozoan cell is preceded by nuclear division. The nuclei of Protozoa may be diploid, possessing two sets of chromosomes, or haploid with only one set of chromosomes. Normally both types of nuclei divide mitotically to give daughter nuclei of similar type. Polyploid nuclei also occur, for example in *Amoeba proteus*, and division here is more complex. The nucleus may split into a number of smaller nuclei each with a complete set of chromosomes.

(i) Types of Binary Fission

The division plane in protozoan cells varies among the various protozoan groups. In naked amoebae binary fission is at its simplest since there is no definable division plane. The cell may round up and division of the cytoplasm into two more or less equal daughter amoebae occurs. In some genera the process of plasmotomy occurs, where multinucleate individuals divide into two or more daughter cells each with several nuclei. There may not be nuclear division at this stage, instead mitosis may be found at any time in such organisms. Plasmotomy is characteristic of larger naked amoebae species, e.g. *Pelomyxa*.

Among test or shell-bearing amoebae the process of cell division is of necessity more complex because skeletal structures must also be replicated. Among proteinaceous shelled species two types of test are found, one composed of numerous alveoli and the other with the appearance of a smooth continuous coat. Netzel (1975a, b, c, 1976) has described the process of alveoli formation in *Centrapyxis discoides* and *Arcella vulgaris*, where daughter alveoli are formed in the cytoplasm of the cell prior to division. During division pseudopodia sheets are extended from the aperture of the parent cell forming a chamber inside which the alveoli are ordered into a replica of the parent shell. When the new shell is complete, the cytoplasmic sheet is retracted and binary fission occurs. Siliceous tests are composed of oval or circular plates which are usually formed prior to division and stored in the cytoplasm of the parent. During division the cytoplasm is extruded from the aperture of the parent test and the reserve shell plates are transferred from the parent cytoplasm to a position around the cytoplasmic extrusion. The extruded cytoplasm is strengthened by a central core of microtubules, and the shell plates are held at the end of finger-like processes by adhesion plaques of concentrated microfilaments (Ogden and Hedley, 1980; Figure 3.1).

Foraminiferan reproduction is complex, and it is now the generally accepted view that alternation of sexual and asexual generations occurs in the group, though there may be exceptions to the general pattern (Grell, 1967). Typically a diploid agamont, in the asexual phase, divides many times, including a meiotic division, to form many gamonts which are liberated as tiny amoeba-like organisms, each of which will secrete a shell around itself and grow into a larger haploid gamont. When fully grown the gamonts produce many isogametes which bear two unequal flagella. These gametes are liberated into the sea where they undergo fertilisation to produce zygotes. The zygotes eventually give rise to diploid agamonts, which proceed to secrete a shell and undergo growth before repeating the cycle (Figure 3.2). There are, however, some interesting variations on this basic life-cycle pattern. In *Rotaliella heterocaryotica* autogamy occurs in the gamonts, and in other species, e.g. *Glabratella sulcata*, the gametes are not released into the sea; instead two gamonts come together and the gametes are exchanged and unite within the space formed by the two shells.

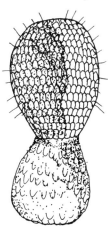

Figure 3.1: Division in *Euglypha*. The extruded daughter contains preformed scales and remains attached to the parent until the scales are arranged into an ordered test.

The radiolarians divide in a number of ways. Some undergo cellular division, but others carry out multiple fission leading to the production of 'crystal swarmers'. Such species are difficult to study in culture because 'swarmers' tend not to develop further (Anderson, 1980). 'Swarmers' may be flagellated and usually contain crystals and abundant stores of lipid, which may be derived from the parent, though this is not certain.

Most flagellates undergo binary fission in a longitudinal plane (Figure 3.3), though in some dinoflagellates, e.g. *Ceratium*, the division plane is oblique and in others, e.g. *Oxyrrhis*, the fission plane is transverse. The usual fission process into two daughter cells involves division of the flagellum and flagellar structures, i.e. the blepharoplast and chromatophores. Simultaneously the nucleus undergoes a mitotic division and the nuclei produced migrate laterally as the cytoplasm splits from the anterior to the posterior of the cell. Armoured dino-flagellates deal with the problem of replicating the theca during division in several ways. In some cases one of the progeny retains the parent theca and the daughter develops a new one, or the old theca may be abandoned by both progeny, each forming a new theca. Although binary fission is the norm in flagellates, some species undergo multiple fission and some show budding.

Figure 3.2: Life-cycle of the Foraminiferan *Elphidium*. a — diploid agamont, b — gamont formation, c — young gamont, d — haploid gamont, e — fertilisation, f — young agamont.

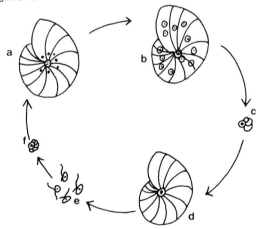

Source: Based on Grell (1967).

Ciliated Protozoa undergo binary fission in a transverse plane (Figure 3.4), but there are exceptions. *Colpoda cucullus*, for example, employs reproductive cysts during division and the suctorians have evolved a process of budding, producing young which are morphologically dis-similar to their parents. The presence of complex body or oral ciliary structures requires replication of kinetosomes (the basal body of the cilium) during the reproductive process. The replication of somatic or body ciliature begins with an increase in the number of kinetosomes in the region where the division plane is destined to occur. These kineto-somes will ultimately give rise to the cilia of the posterior region of the anterior daughter cell (the proter) and the anterior cilia of the posterior daughter (the opisthe). The replication of the oral ciliature, a process termed stomatogenesis, varies within the various ciliate groups, but four

Figure 3.3: Binary Fission in *Euglena*, Showing Division of Flagella and Flagellar Structures, Nucleus and Finally Longitudinal Division of the Cytoplasm.

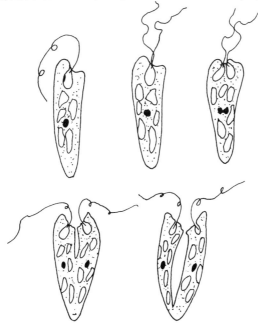

Figure 3.4: Binary Fission in *Tetrahymena*. a: The cell prior to division. b: The beginning of division showing the area of developing kinetosomes to the left of the stomatogenic kinety. c: The oral ciliature of the opisthe is developed and the macronucleus can be seen to undergo division. d: Prior to final stage of division. dk — developing kinetosomes, sk — stomatogenic kinety, m — macronucleus, cv — contractile vacuole, oc — oral ciliature.

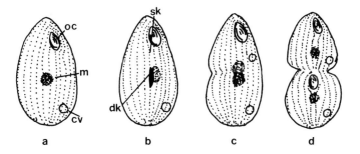

a   b   c   d

broad categories are recognisable. In the more primitive ciliates, where specialised oral ciliature is lacking, telokinetal stomatogenesis occurs. This involves transection of somatic kinetosomes, with those on the anterior end of the opisthe producing a new oral area. The majority of spirotrichs and hymenostomes typically show parakinetal stomatogenesis during which an apparently randomly organised field of cilia-free kinetosomes develops from or alongside the stomatogenic kinetics (i.e. somatic kinety of the central cell region), subsequently becoming properly organised to form the oral ciliature of the opisthe (Figure 3.4). Some hymenostomes, spirotrichs and the peritrichs exhibit bucco-kinetal stomatogenesis in which some of the kinetosomes involved in the oral ciliature development of the opisthe are derived from the oral ciliature of the parent. Apokinetal stomatogenesis also occurs in some spirotrichs. Here a field of kinetosomes appears with no apparent parental somatic or oral kinetosomal origins (Curds et al., 1983). Thus the anterior daughter retains the parent mouth and the posterior daughter becomes the possessor of the newly formed cytostome.

In *Tetrahymena*, which has been extensively studied, the micronuclei and macronucleus behave in distinctly different manners during division. First the micronuclei undergo mitotic division, and when this is nearly completed, the macronucleus divides amitotically, elongating and then becoming constricted centrally, forming two daughter macronuclei. Meanwhile the daughter micronuclei have migrated anteriorly into the proter and posteriorly into the opisthe. Upon the termination of macronuclear division the cytoplasm quickly shows a transverse furrow with rapid division of the parent cell into two identical daughter cells. The original contractile vacuole is retained by the opisthe and a new contractile vacuole is formed by the proter while the micronuclei migrate. On formation osmoregulation is commenced immediately by the new contractile vacuole (Figure 3.4).

Where reproductive cysts occur, as in *Colpoda cucullus*, the cell rounds up and a thin walled cyst is formed, which differs from a resistant cyst in wall thickness and the ability to withstand adverse conditions. Within the cyst the micronucleus divides mitotically, followed by division of the macronucleus. The cytoplasm then divides producing two daughter cells within the cyst wall. Each daughter cell undergoes a similar division cycle so that four daughter cells are contained within the cyst. The cyst wall then breaks down liberating the four progeny (Figure 3.5).

The Suctoria, which are sedentary and lack cilia in the mature stage, have of necessity developed an alternative to binary fission. The production of similar non-motile progeny would lead rapidly to over-crowding and increased competition for food as well as space. In common with many metazoans, the suctorians produce a motile dispersal immature stage, which is sometimes referred to as a swarmer. The young are produced one at a time by a process of endogenous or exogenous budding. Endogenous budding is particularly interesting, and the rapid evagination of the swarmer is a spectacular event to observe microscopically. Ultrastructural studies show that the development of the embryonic cavity during endogenous budding involves a sequence of changes in the cortex of the parent. The embryonic cavity is initiated in proximity to a single subcortical kinetosome which replicates by the production of a pro-kinetosome. The replication continues until a saucer-shaped field of kinetosomes is formed just beneath the cortex, usually in the half of the body opposite the stalk. At an early stage of this development the cilia become organised in rows. Before ciliogenesis each kinetosome is associated with a kinetodesmal fibre, a single post-ciliary microtubule and basal microtubules, all of which become part of the infraciliature of the swarmer. (For details of the infraciliature structure in ciliates consult Chapter 4, Section B.) During the replication of kinetosomes the embryonic cavity continues to expand and its wall is characterised by an increase in the number of subcortical microtubules. At a later stage micronuclei are observed in close proximity to the wall of the cavity. The division of the macronucleus correlates with the budding process (Curry and Butler, 1982).

The first visible sign of budding is when the embryonic cavity has completed its development and there is breakdown of material from what becomes the birth pore, causing a small protruberance on the parent cell (Figure 3.6a). This enlarges and is accompanied by strong cytoplasmic movements. The walls of the cavity then emerge quickly from the pore to become the cell wall of the swarmer, exposing the cilia (Figure 3.6b). The swarmer remains attached to the parent by a thin cytoplasmic bridge which eventually breaks, liberating the young non-feeding ciliated individual (Figure 3.6c). It is suggested that the evagination process occurs by the removal of subcortical microtubular structures in the region of the birth pore, which prevents the cell wall holding back against the hydrostatic pressure of the cytoplasm (Curry and Butler, 1982). Other workers (Henk and Paulin, 1977) have observed a reduction in the number of subcortical microtubules in reproducing adults compared with non-budding individuals, which lends weight to the mechanism proposed by Curry and Butler (1982).

The swarmer usually metomorphoses within about 30 minutes of release from the parent. It settles on a substrate with the ventral surface downwards. The cilia cease beating, the stalk is secreted by the scopuloid and tentacles are then produced from tentacle primordia of microtubules arrays. Simultaneously the cilia shorten and gradually disappear.

Figure 3.5: Asexual Reproduction in *Colpoda cucullus* by Means of Reproductive Cysts.

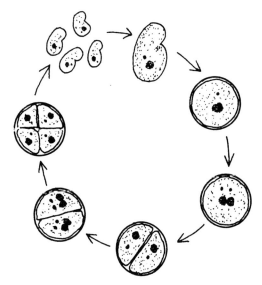

Within the Protozoa and Animalia asexual reproduction can take various forms. Sibly and Calow (1982) have attempted to analyse the diversity of asexual reproductive patterns in relation to the ecology and mode of life of various organisms. The aim of these authors was to formulate models of reproduction and to predict which pattern is optimal, and hence adaptive, under particular ecological conditions. In the Protozoa multiple fission has evolved most commonly in parasitic species, although it does occur in some free-living sarcodine taxa. Binary fission is, however, the most usual mode of asexual reproduction among free-living species. The implication, although unproved, is that the growth conditions are better for parasitic species, which have evolved greater reproductive potential in multiple fission. The argument is that under conditions which are good for growth, assuming a constant mortality, more smaller offspring should be produced. Optimising reproductive output is of course advantageous to a parasitic species because it enhances the chance of finding new hosts.

Figure 3.6: Endogenous Budding in *Podophrya*. a: The swarmer can be seen as a dense area inside the parent and a slight protruberance appears on the cell surface indicating the position of the birth pore. b: The swarmer is evaginated and remains attached to the parent for a few minutes. c: The swarmer is liberated.

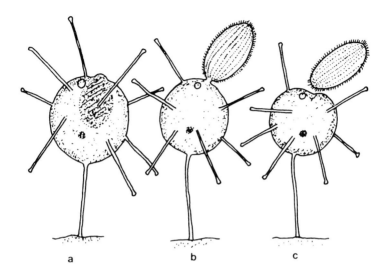

a   b   c

For sessile organisms employing asexual reproduction the optimum mode is attachment to the parent throughout development. Budding as seen in the suctorians and Cnidaria is the form of reproduction one would expect to result from a sessile life-style, where the parent's normal functions, such as feeding or performing protective contractile movements, are not impaired by offspring attachment. Where attached offspring would be an impediment to function, as in sessile Anthozoa, laceration and longitudinal fission are common (Sibly and Calow, 1982).

(ii) Mean Cell Volume Variation

The size attained by a protozoan cell before binary fission commences is not fixed for a given species, and varies as a function of environmental and biological factors. This simple fact can be demonstrated experimentally by growing Protozoa in culture without replenishing the food supply. The ciliate *Colpidium campylum* grown in baked lettuce medium (Figure 3.7) shows an initial high population mean cell volume which coincides with plentiful bacterial food supply and relatively low ciliate density. During the progressive increase in ciliate population density, bacterial food supply decreases and mean cell volume declines. As the food supply becomes exhausted the cells divide at a smaller size, so that there comes a point when population biomass ceases to increase but is broken up by division into more units or cells. Arguably this is a characteristic of laboratory culture conditions, but in all probability wild populations wax and wane in the same manner, as local pockets of bacterial species palatable to particular protozoans flourish and die. The production of a large number of individuals as food supply diminishes has clear ecological advantages, since it enhances the chances of an individual finding another suitable food source and thus perpetuating the species.

Figure 3.7: Mean Cell Volume Variation and Protozoan Cell Density in Response to Bacterial Concentration. ● – mean cell volume, ■ – protozoan density, ▲ – bacterial density.

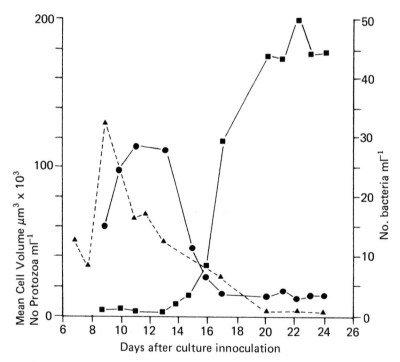

Variation in the mean cell size of populations of Protozoa was noted as early as 1937 by Harding in the ciliate *Glaucoma*. More recently attempts have been made to elucidate the factors responsible for mean cell volume variation in a number of protozoan species (James and Read, 1957; Hamilton and Preslan, 1970; Curds and Cockburn, 1971; Lee and Fenchel, 1972; Laybourn, 1975; Finlay, 1977; Laybourn-Parry et al., 1980). Food concentrations and hence food consumption appear to be major factors responsible in some ciliate species. Under steady-state conditions, that is in a chemostat, the population density of the ciliate cells also exerts an influence. Here an increase in population density causes an increase in mean cell volume (Hamilton and Preslan, 1970). Temperature has been implicated as a major controlling factor in some species. Lee and Fenchel (1972) showed that increasing temperature induced a decrease in mean cell volume in *Euplotes balteatus*. The response to temperature however varies: *Vorticella microstoma* and *Spirostomum teres* respond in the same way as *E. balteatus*, whereas *Tetrahymena pyriformis*, *Paramecium aurelia* and *Frontoria leucas* increase mean cell volume with increased temperature (Laybourn and Finlay, 1976). The differences in response may be related to the temperature tolerances of the species concerned.

Most of the investigations on this problem have focused on ciliates and little work has been carried out on flagellates or amoebae. However, Rogerson (1981) has demonstrated the phenomenon of mean cell volume variation in *Amoeba proteus*. In this large amoeba food concentration affected mean cell volume, but in addition temperature appeared to exercise a direct effect, so that at any given food concentration cell size varied with temperature. Two species of small amoebae, *Saccamoeba limax* and *Vannella* spp. showed variation in mean cell volume related to temperature, but in this study the food concentration effect was not considered (Laybourn-Parry et al., 1980). In common with the ciliates which have been studied, these small amoebae responded differently to temperature – in *Saccamoeba* maximum size occurred at 15°C, but *Vannella* showed maximum cell size at 20°C.

Clearly the mechanisms controlling the average cell size at which division is initiated are complex and probably vary from one species to another. In some protozoans temperature alone appears to be the major governing factor, while in others food concentration and temperature act together. Salt (1975) has put forward an hypothesis which in part explains the variations seen in some species. He investigated the ciliate predator *Didinium nasutum* and found that mean cell volume increased relative to population density provided food (the ciliate *Paramecium*) was replenished. When food was not replenished mean cell volume decreased. The hypothesis rests on the facts that food intake per individual per generation declines as the population density increases, but meanwhile the metabolic and respiratory rates of individual protozoans decline at an even faster rate, which results in a surplus of food or energy being ingested and stored despite declining food intake. Hence there is increased cell volume, where food supply is plentiful and in excess of the protozoan population requirements. When the food supply declines there is some stored endogenous energy to sustain the population before mean cell volume starts to decrease.

Variation in the mean cell volume of a given species population has a number of important implications in protozoological studies. The phenomenon reduces the value of size as a criterion in taxonomic studies, but more importantly it means that physiological studies on growth and division rates must also include a consideration of mean cell volume. This fact is even more significant in respiratory studies where meaningful results are only obtained if metabolism is related to some parameter of size.

(iii) Growth and Factors Influencing Growth and Division

The rate at which Protozoa increase cell size, or grow, and the rate at which they divide, are governed by a complex range of environmental and biological factors. Evidence suggests that the mechanism controlling the rate of reproduction or binary fission may change more slowly in response to changing environmental conditions than the growth controlling mechanism (Kimball et al., 1959; Hamilton and Preslan, 1970). Thus when temperature decreases, or when food supply declines, the growth rate quickly slows down whereas the reproductive rate responds more slowly; hence the production of a population of individuals with an overall lower mean cell volume.

Temperature. Temperature is a fundamental factor influencing the metabolic functions of cold-blooded organisms. Although many species are able to exercise some control and compensation over their metabolic rates, the temperature regimes experienced in the habitat occupied by an animal have a profound effect on almost all aspects of physiological functioning. Since the majority of protozoa are aquatic they live in a medium which undergoes gradual temperature changes throughout the annual seasonal cycle. The intertidal and terrestrial environments are subject to rapid and often dramatic temperature fluctuations. Protozoa which live in intertidal and terrestrial communities, however, live in the interstitial waters of sediment and soil where they are less subject to sudden temperature changes. There have, of course, been many studies on the effects of temperature on the growth and reproduction of Protozoa. Inevitably, the ease with which Protozoa can be cultured and established as laboratory organisms has resulted in many physiological investigations being performed at temperatures outside the range normally experienced in the natural environment. Although intrinsically interesting, the information provided by such studies does not give us a great deal of insight into the functioning of these creatures in their natural habitat, or their interrelations with other organisms in the food web. More recently an ecological approach has been adopted by some physiologists, in an attempt to elucidate the reproductive potential of Protozoa in the context of the normal environmental regimes experienced in the wild.

The effect of temperature on the rates of growth and reproduction can be demonstrated experimentally. The ciliates *Stentor coeruleus* and *Colpidium campylum*, for example, show a typical increase in the rates of growth and division, thus decreasing generation time, as temperature increases within the environmental temperature range. *C. campylum* increases growth rate by more than three times (Q_{10} 3.40) between 10°C and 20°C (Laybourn and Stewart, 1975a; Laybourn, 1976b). *Amoeba proteus* shows much the same pattern of growth related to temperature as manifested by ciliates. Within the temperature range 10-20°C growth rate was highest at 20°C. Furthermore, in *A. proteus* the largest cells developed at 10°C where growth was lowest, while at 20°C mean cell volume was only half that at 10°C (Rogerson, 1981). The medium-sized amoeba *Polychaos fasiculatum*, feeding on bacteria, and small limax amoebae also increase growth with increasing temperature between 5°C and 25°C. Specific growth rates rise from 0.933 ± 0.18 up to 12.9 ± 0.95 (Baldock and Baker, 1980). *P. fasiculatum* is probably near its thermal limit at 5°C because generation times are extremely long, around 323 hours compared with 23.5 hours at 25°C. The small amoeba *Saccamoeba limax* would appear to have a narrower temperature range because growth decreases above 20°C and generation times increase from 4.03 hours at 20°C to 5.98 hours at 25°C. At the other end of the temperature scale the comparatively long generation times at 10°C suggest that the lower thermal limit is being approached (Baldock *et al.*, 1980). Certain species do not grow at all at low temperatures and are clearly thermophilic, for example *Vexillifera baccillipedes*, which will not grow at 10°C or below (Baldock *et al.*, 1980).

The above examples highlight the need for a wide, detailed approach to the problem of quantifying and understanding the metabolic responses of protozoans to environmental factors, because superficially what may appear to be a similar response, on detailed examination is often the result of dissimilar functional energetic responses within the organism.

The question of whether Protozoa in the extreme environments of the polar regions are able to perform temperature compensation, that is achieve higher rates of growth at lower temperatures than Protozoa from warmer latitudes, has been considered by Lee and Fenchel (1972). They studied three species of the ciliate *Euplotes* from the antarctic (*Euplotes antarcticus*), from Denmark (*Euplotes vannus*) and from Florida (*Euplotes balteatus*). With careful gradual acclimatisation all three species could be made to survive at temperatures outside their normal temperature ranges. *Euplotes balteatus* (the tropical species), for example, quickly became abnormal and died if suddenly transferred from $21°C$ to temperatures below $10°C$ or above $31°C$. However, with slow acclimatisation this species could be induced to divide indefinitely at $8°C$ and $33.5°C$. The other species, *E. antarcticus* and *E. vannus*, behaved in a similar fashion but within different temperature ranges. A Q_{10} of 1 or slightly less in the temperature range 5-$10°C$ was demonstrated in *E. antarcticus*, maximum growth being achieved at $5°C$. Compensation did not, therefore, occur in the antarctic species of *Euplotes*; its growth rate may be found by extrapolating the growth rates of other species studied at low temperatures. Lee and Fenchel make the point that while morphologically identical ciliates, with few exceptions, are worldwide in distribution, physiological forms or races occur in different climates. In all probability this characteristic is not restricted to ciliated Protozoa, but may also be true of other free-living groups.

A definable relationship between cell size and reproductive rate has been demonstrated in ciliated Protozoa (Fenchel, 1968; Finlay, 1977) and in amoebae by Baldock *et al.* (1980). When generation times of a large number of Protozoa are related to mean cell volumes at a given temperature, it becomes very clear that there is a positive relationship between log T (generation time) and log V (cell volume) at any given temperature. The relationship is shown for ciliates in Figure 3.8. Fenchel (1968) found that all the species he included in his study showed increased cell size at lower temperatures; a trend that was particularly pronounced in some species, for example *Uronema*. Both Baldock *et al.* (1980) and Finlay (1977), however, found more variability in cell size with temperature, some species decreasing rather than increasing cell volume at lower temperatures. The generation time of any ciliate can be derived from the type of regression shown in Figure

3.8, provided one knows the cell volume of the species population, which is easily measured. Given the large number of ciliate species one finds in natural protozoan communities, and the impossibility of measuring growth and reproduction in each species directly, a rapid method of estimating the reproductive potential of a wide range of species in a habitat is invaluable, and provides the physiological ecologist with a method of calculating the potential production of a community. From the generation time (T), the intrinsic rate of natural increase rm = 1n 2/T in animals dividing by binary fission, and the finite rate of increase, i.e. the factor by which a population will multiply per unit time (= λ), can be calculated.

Figure 3.8: The Relationship between Reproductive Rates Statistics (T, r_m and λ) and Cell Size in a Range of Ciliates at 20°C. Key to species: ● — *Stentor polymorphus*, ○ — *Paramecium aurelia*, ▼ — *Paramecium bursaria*, ▽ — *Spirostomum teres*, ■ — *Colpidium campylum*, □ — *Loxocephalus plagius*, △ — *Vorticella microstoma*, ◕ — *Tetrahymena pyriformis*, ▲ — *Chilodonella uncinata*, ◨ — *Cyclidium glaucoma*.

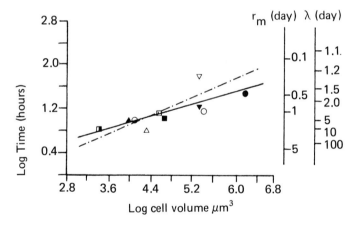

Source: Broken line from Fenchel (1968). Redrawn from Finlay (1977).

Food. Since energy is an essential commodity to all heterotrophic animals, any variation in the quality or quantity of the energy source or food supply may have a profound effect on the growth and reproductive physiology of an organism. The concentration of the food supply can influence the growth and reproductive potential of Protozoa, particularly those species which voraciously consume bacteria.

Generally bacterivore species show increasing growth and reproductive rates as food concentration increases, so that below a critical prey density ingestion of energy is insufficient to achieve maximum production, while above the critical food concentration growth becomes independent of available energy (Proper and Garver, 1966; Curds and Cockburn, 1968; Laybourn and Stewart, 1975; Laybourn, 1976a; Taylor, 1978a). Figure 3.9 shows the typical growth curves related to food supply for a number of ciliates. Temperature has the effect of raising the growth rate at any given prey concentration in many species – Figure 3.10 illustrates the phenomenon in *Colpidium campylum*. The growth curves of *Cyclidium glaucoma* and *Colpodium colpoda* (Figure 3.9) vary as a function of each species' ability to exploit the bacterial food organism, in this case *Aerobacter aerogenes*. Since all these species were investigated separately, the question of competition does not arise, but in the wild a species such as *Colpidium colpoda*, which can successfully exploit lower food concentrations more effectively than other species, would have a positive competitive advantage.

Probably, growth is limited at high food concentrations by the rate at which ingested bacteria can be processed rather than by the rate at which bacteria can be ingested. *Colpidium campylum* continues to increase consumption after maximum growth levels have been achieved (Laybourn and Stewart, 1975), which is clearly a wasteful practice in energetic terms. Over-consumption in abundant food conditions resulting in poor digestive efficiency is not uncommon among metazoans, and is well documented in filter-feeding Cladocera; it is not inconceivable that a similar phenomenon occurs in Protozoa when a rich food supply becomes available.

The growth of carnivorous Protozoa, i.e. those feeding on other Protozoa and small metazoans, responds to increasing food concentration in a different manner. *Amoeba proteus* preying on *Tetrahymena pyriformis* showed reduced production where prey density and consumption increased (Rogerson, 1981). At 20°C maximum growth was achieved where 2.2 *Tetrahymena* were captured per hour, and at 10°C where 0.7 *Tetrahymena* were ingested per hour. Higher levels of prey ingestion resulted in a decline in the growth rate. The sedentary suctorian predator *Podophrya fixa* also shows a decline in growth at higher prey densities, but in this species prey capture and growth show a corresponding decline (Laybourn, 1976a). *Stentor coeruleus*, a large omnivorous ciliate, when fed experimentally on *Tetrahymena*, undergoes a reduction in growth at higher prey densities at 15°C, but not the higher temperature of 20°C. Moreover at higher food concentrations food consumption continues to increase despite reduced growth (Laybourn, 1976b).

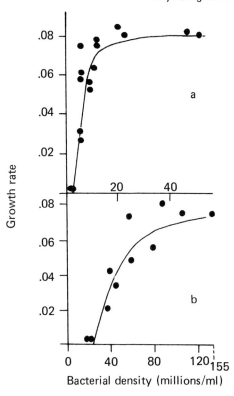

Figure 3.9: The Growth Rates of a — *Colpidium colpoda*, b — *Cyclidium glaucoma* in Response to Bacterial or Food Density. Growth rate calculated as $(\log_e N_t - \log_e N_0)/t$ where N_0 and N_t are the initial and final number of ciliates and t is the elapsed time.

Source: Taylor (1978a), with the permission of Springer-Verlag, New York.

Figure 3.10: The Growth of the Ciliate *Colpidium campylum* in Relation to Temperature and Food Supply. $\bigcirc - 10^{\circ}C$, $\blacktriangle - 15^{\circ}C$, $\blacksquare - 20^{\circ}C$.

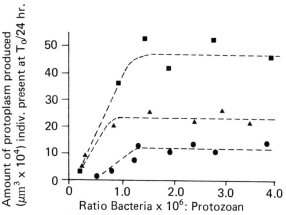

Source: Laybourn and Stewart (1975), with the permission of the British Ecological Society.

The type of food available and the potential energy content of the food may also result in variations in the growth and reproductive rates. In the natural environment the preferred ideal food may not be available, and organisms must survive on less suitable, often less nutritive foods. Among exclusively bacterial feeding Protozoa, the species of bacteria consumed may determine the growth and reproductive rates achieved. Curds and Vandyke (1966) demonstrated that while some bacterial species supported good reproductive rate in a given ciliate, other bacteria would only support poor reproduction, while some bacteria were positively toxic to the ciliate. *Arcella vulgaris*, a testate amoeba, survived and reproduced very slowly on diets of fungi and algae, or a mixture of both, but when fed on bacteria or diets containing bacteria with algae or fungi, the rate of reproduction increased two- to three-fold (Laybourn and Whymant, 1980).

Moisture. Soil-dwelling Protozoa face the prospect of desiccation of their environment periodically, usually in a seasonal cycle. Lousier (1974) investigated the response of testate amoebae to soil moisture in an aspen wood in Alberta, Canada. Increased soil moisture resulted in a significant increase in active Testacea in the soil except in the top 2 centimetres, and a decrease in encysted forms. Furthermore, the amount of growth achieved or production was higher and generation times shorter in the watered plots. The reproductive rates of testate amoebae are low in the field, estimates of the order of ten generations a year having been suggested for testate inhabitants of temperate bogs and fens (Heal, 1964), although higher rates are achieved under laboratory conditions, where the generation time may be around eight days. The thickness of the test and its size may be restricted by the thickness of the water film in drier sites. The test size of *Nebela* can be increased by flooding the *Sphagnum* moss in which they live. Generally, species with larger pyriform tests are restricted to *Sphagnum* submerged pools, while *Sphagnum* above the water level is inhabited by species with small or flattened tests (Heal, 1963).

Light. Light can be an important factor influencing the reproductive cycle of green flagellate species. *Chlamydomonas moewusii* appears to divide at the transition between light and dark phases (Scherbaum and Loefer, 1964). In some species of *Euglena* mitosis appears to be confined to the dark period of the diurnal cycle. The first division stages appear about two hours after the onset of darkness, and between half an hour to two and a half hours later maximum cell division occurs (Leedale, 1967). In the annual cycle of primary production by the phytoplankton, light plays an important role. During the winter months production by the autotrophic Protozoa and other unicellular organisms in the phytoplankton is extremely low; this is the result of low available incident radiation. In the spring the levels of radiant energy available for primary production increase and stimulate the so-called spring bloom or vernal bloom of the phytoplankton. The spring bloom involves rapid growth and division following roughly an exponential course over a few weeks. In most aquatic environments the spring outburst of the phytoplankton is short-lived and a decline caused by a variety of factors brings about a cessation in production. Nutrient depletion and zooplankton grazing are responsible for the decline in production, their importance varying according to the type of aquatic habitat. During the summer production continues, but at a lower level. An autumn blooming also occurs in some aquatic environments, particularly stratified lakes undergoing an autumnal overturn which releases nutrients from lower waters.

Obviously in any given protozoan species there is an array of factors which act together in influencing not only the rate of growth, but also the rate of asexual reproduction. Temperature is a major overriding factor, particularly in heterotrophic Protozoa, often influencing the magnitude of a response to other factors such as food concentration. Inevitably because temperature controls metabolism and activity it may alter an organism's ability to capture and ingest food as well as its ability to respond to changes in environmental pH, light, moisture, etc. A clear understanding of the complexity of growth and reproduction in the normal asexual life-cycle of free-living Protozoa can only be gained from detailed energetics studies, as some of the foregoing examples highlight. The simplistic view many biologists hold of protozoan growth physiology is incorrect. We are not dealing simply with small pieces of protoplasm which grow, reach a predestined size and divide into two to repeat the cycle; instead the situation is highly complex. The size attained before division occurs is determined by the environment and biological characteristics of the habitat, and the growth rates and division rates are similarly variable and influenced by outside factors.

C. Sexual Reproduction

Sexual reproduction is a normal phase in the life-cycle of many of the parasitic Protozoa, alternating with an asexual phase. In all probability this is the result of adaptation to a parasitic life-style in a two-host life-cycle. Among free-living Protozoa sexual reproduction is a rarer event, and usually only resorted to in times of adversity, when food becomes depleted or when environmental conditions change radically. The different forms of sexual reproduction displayed by Protozoa are essentially a means of genetic reorganisation which does not necessarily result in an increase in numbers.

Among the ciliated Protozoa the ability to reproduce by a sexual process is almost universal. Ciliates use a procedure known as conjugation for the exchange of genetic material; a process which does not produce an immediate increase in numbers. In the absence of suitable conjugatory partners some ciliates resort to the process of autogamy. Although sexual processes occur in the Sarcodina, sexual reproduction is not a characteristic of all members of the group. The foraminiferans alternate a sexual phase with an asexual phase (see Section B(i)) and some helizoans reproduce sexually (Grell, 1967). Sexual competence is not widespread among free-living flagellates, and is restricted to the orders Chloromonadida, Dinoflagellida, Prasinomonadida, Prymnesiida and Volvocida.

Isogamous (gametes alike) and anisogamous (gametes dissimilar) reproduction is found in free-living protozoan species. Gametes, whether isogamous or anisogamous, develop from the typical asexual vegetative forms. Anisogamous gametes are designated microgametes and macrogametes and are analagous to the sperm and ova of metazoans. The microgametes are small, mobile and often numerous, whereas the macrogametes are large, immobile and usually produced in lower numbers.

(i) Factors Stimulating Sexual Reproduction

Sonneborn (1957, quoted in Kudo, 1971) writes 'In the absence of sexual processes paramecia grow old and either die or become genetically extinct by losing their micronuclei. Survival depends upon the occurrence of fertilization before senility has gone too far. The fertilized animals are reinvigorated and start new young clones.' The statement summarises the major need and cause of a sexual cycle in many ciliates. As a result of conjugation or autogamy the old macronucleus is replaced by material derived from the fusion of micronuclei from different animals. Thus the cell is in essence physiologically rejuvenated by the development of a new macronucleus. The ageing of ciliates has been studied in culture, although it must be noted that ageing is probably a less common phenomenon in the wild. Nonetheless in natural communities a species may develop rapidly in response to a rich food supply and/or high temperatures, and then progress through a decline towards senescence at which point a sexual phase may be initiated.

The ability of individuals to conjugate changes during cultivation of a clone in the laboratory. During the first stage a clone is immature and neither conjugation nor autogamy occurs. The immature period varies from one species to another, lasting about three to five months. A clone then enters a mature phase within which normal conjugation can occur. The mature period can vary from as little as three days to a month or more as in *Paramecium aurelia*, and in some cases several years, e.g. *P. bursaria*. 'Maturity' is succeeded by senescence, where conjugation ceases but autogamy occurs, though becoming progressively prone to abnormalities. Ultimately death of the clone ensues (Raikov, 1972). Whether the pattern is exactly the same in nature remains to be seen.

Ageing may also be a contributory factor in those flagellates capable of sexual reproduction. Healthy, rapidly-growing cultures are in most cases not sexually competent, and it is only in the later stages of culture growth that sexual competence arises (Coleman, 1980). Sexual reproduction has been observed in collections of flagellates taken from the wild on many occasions.

Ageing or senescence appears to be an important triggering factor in sexual competence, but there is evidence that some environmental factor may be responsible for inducing conjugation, certainly in *Stentor*. Two cultures derived from widely separated habitats have been shown to undergo conjugation simultaneously at room temperature, and even more remarkable conjugation was noted on the same day in cultures maintained at ambient temperature but separated by a distance of 30 miles (Rapport *et al.*, 1976). Experiments with *Stentor* show that conjugation tends to occur above $20^{\circ}C$ and is associated with a light cycle. Apparently environmental shock or starvation are not necessary prerequisites, as had previously been assumed. A closely-related species, *Blepharisma*, also tends towards conjugation at higher temperatures (Bleyman, 1975). The fact that conjugation often occurs in wild populations during some definite season in the year, for example *Nassula ornata* in May (Raikov, 1972) and *Loxodes striatus* in August (Bogdanowicz, 1930), suggests that temperature may play an important role in initiating the sexual process, at least in some ciliates.

(ii) Isogamous Reproduction

Flagellated Protozoa capable of sexual reproduction may be isogamous or anisogamous. Examples of isogamous species are *Chlamydomonas* and *Gonium*. Coleman (1980) has described the sequence of events in detail. The first stage in colonial forms is mating-type-specific flagellar agglutination reaction, where the flagellar tips of sexually competent cells are capable of sticking together resulting in the clumping of colonies in large aggregates. At the same time the colony matrix begins to autolyse. There are a number of distinct stages in the reproductive process after sexual competence has occurred.

The first stage is pairing, which may commence a few minutes after the colonies are mixed. The flagellar tips of pairs of gametes become firmly anchored and bob vigorously against this anchoring force, thus leading to contact between the cell apices and the instantaneous formation of intracellular cytoplasmic bridges. This next stage is fusion and the bridge is formed by a tiny protoplasmic protrusion from the subapical region of one gamete, making contact with and fusing with a similar protrusion on the other gamete. Once the apical bridge has been formed fusion is completed within 1-2 minutes. The zygote, thus formed, then undergoes a period of growth in which carbohydrate and oil accumulate. A heavy wall is laid down around the zygote and the two nuclei derived from the two gametes fuse.

The sexual reproductive processes carried out by ciliates are essentially isogamous. Where two individuals are involved the process is described as conjugation, which is a special type of nuclear reorganisation peculiar to ciliates. The function of conjugation is genetic recombination and is the process characterised by the temporary mating of two individuals. Conjugation in *Paramecium caudatum* has been investigated in considerable depth and the information reviewed by Raikov (1972). The process is similar in other species so that the stages shown by *Paramecium* can be used to illustrate the phenomenon of conjugation (Figure 3.11).

First two conjugants from compatible mating strains come together; soon after, the micronuclei of each individual enter a characteristic crescentic form of the first maturation prophase and then complete the division. The daughter micronuclei of each conjugant than undergo a second meiotic division. Of the eight micronuclei produced, seven break down, leaving the surviving micronucleus to undergo a third maturation division. This division, which is equatorial, results in two spindle-shaped pronuclei. One of the pronuclei of each conjugant remains stationary while the other penetrates the joined pellicles of the conjugants passing into the partner. Thus an exchange of pronuclei is achieved. In each conjugant the newly-entered pronucleus approaches the stationary pronucleus and the two fuse to form a synkaryon, which then undergoes two mitotic divisions. The conjugants usually separate after the first of these mitotic divisions, so that the second division occurs in the exconjugants. The four derivatives of these two divisions are situated two in the anterior half of the cell and two in the posterior half. The two anterior nuclei increase in size and become the macronuclear anlagen.

Figure **3.11**: Conjugation in *Paramecium aurelia*. a-b: Preconjugation interaction. b-d: Eight haploid micronuclei are produced by two successive meiotic divisions. d-e: Seven of the nuclei disintegrate leaving only one. f: Nucleus divides to produce two nuclei, one of which is stationary, the other migratory. g: Exchange of migratory gametic nuclei. h: Fusion of exchanged migratory nuclei and stationary nucleus to form a synkaryon. j-k: Synkaryon divides twice to produce four nuclei. l: Two of the nuclei form the macronuclear anlagen and two are the micronuclei of the cell. m: Cell division with the macronuclear anlagen being distributed between the daughter progeny of the division.

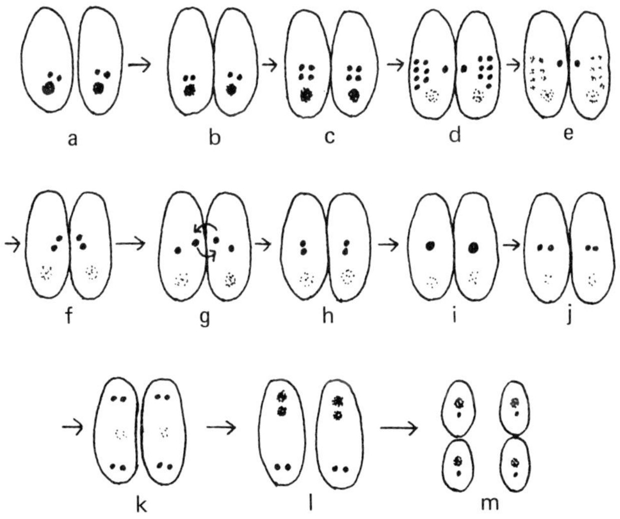

The exconjugant ciliates then divide by binary fission. During this division the micronuclei divide mitotically and the macronuclear anlagen becomes distributed among the daughter cells. Thus the two offspring of the division each have one of the constituents of the macronuclear anlagen. Among the various ciliate taxa which have been investigated the number of synkaryon divisions varies, as does the number of nuclei in the macronuclear anlagen. Conjugation of this temporary variety occurs in almost all ciliates, including all of the Holotricha, the majority of the Heterotrichida and Hypotrichida and all the Entodiniomorphida, but not in the Peritrichia. Some of the sessile Suctoria show total isogamontic conjugation while others practise temporary conjugation between neighbouring individuals. Total conjugation is considered by Raikov (1972) to be more advanced than the temporary conjugation characteristic of most ciliates.

A prerequisite to any sexual encounter between organisms is some form of suitable stimulatory communication, and the Protozoa are no exception. Conjugation is mediated in some ciliates, for example species of *Blepharisma,* by conjugation signals or gamones which are secreted into the medium (Miyake, 1978; Ricci and Esposito, 1981). A typical sequence involves Type I cells excreting a gamone I which gives specific information to Type II cells, which then also excrete a second type of gamone which passes information to Type I cells. The gamones transmit 'chemical information' which causes compatible mating strains or types of one species to undergo transformations facilitating conjugation. Some chemical analysis of gamones in *Blepharisma japonicum* have shown that one of the gamones produced is calcium-3-(2'-formyl-amino-5'-hydroxybenzoyl) lactate (Kubota *et al.*, 1973), which can be synthesised and used to induce conjugation, although synthetic gamones appear only half as active as the natural gamone. Not all ciliates secrete gamones, however; in other species, e.g. *Paramecium*, the conjugatory signals are cell-bound and function as a result of direct contact between two types of cells (Hiwatashi, 1969; Miyake, 1974). Cells of complementary mating strains of *Paramecium* begin ciliary agglutination with each other very soon after they are mixed or come into contact. The mating reaction continues for approximately one hour before the cells unite in conjugatory pairs. The process of conjugation can apparently only occur between compatible mating strains or types. In *Paramecium bursaria* six syngens have been identified, each containing between one and eight mating types. Mating normally occurs between mating types of the same syngen (Sonneborn, 1957). Sixteen varieties or syngens have been identified in *Paramecium caudatum,* each syngen containing two mating types (Hiwatashi, 1949; Sonneborn, 1957).

The process of autogamy, which occurs only in a restricted number of species, involves much the same sequence of events as conjugation, with the major exception of the exchange of pronuclei since it is a function performed by an individual not a pair. *Paramecium aurelia* is typical and in this species both the micronuclei undergo two divisions involving meiosis. Some of the eight micronuclei formed disintegrate, while the others enter a third maturation division, but only one completes the third division to produce two pronuclei. The pronuclei fuse to form a synkaryon which then undergoes two divisions. Of the four derivatives produced by the synkaryon divisions, two become micronuclei and two the macronuclear anlagen. The cell divides once and the macronuclear anlagen is divided between the two daughter cells. The old macronucleus disintegrates during the processes of conjugation and autogamy.

(iii) Anisogamous Reproduction

Some flagellates of the order Volvocida are anisogamous, e.g. *Volvox*, *Eudorina*. All produce microgametes or sperm, which are small biflagellate gametes produced in bundles by multiple mitosis of a mother cell, and large cells or macrogametes, which are described as eggs. When a male colony encounters a female colony the flagella of the sperm bundle appear to stick to the gelatinous surface materials of the female colony (Figure 3.12). Where the sperm bundle becomes anchored to the female colony the female colony matrix dissolves. The sperm bundle then disintegrates into individual sperm which proceed to penetrate the matrix of the female colony, burrowing into the gelatinous material until they make contact with the female gametes, whereupon fusion of gametes takes place (Coleman, 1980).

Conjugation by anisogamous conjugants occurs in some of the sessile peritrich ciliates. Colonies are essentially monoecious, with macroconjugants occurring only on younger colonies and microconjugants developing on older colonies, thus preventing conjugation or 'self-fertilisation' within the same colony. The microconjugants develop by a single equal division of a microzooid in any portion of a colony, whereas the macroconjugants, of which only two differentiate per colony, occur at the top of the two main colony branches. Some of the Suctoria also have distinct anisogamontic conjugation, e.g. *Discophrya* spp. Here the macroconjugants resemble vegetative individuals, while the microconjugants are formed by budding from vegetative parents. The microconjugants are mobile and resemble the dispersal 'larval' stage, but differ in being smaller. On locating a macroconjugant the microconjugant enters the former in a so-called 'inner conjugation' process (Raikov, 1972). This is total conjugation in comparison with the temporary conjugation practised by the majority of ciliates. In suctorians performing this type of total conjugation, two types of budding occur, ordinary vegetative asexual budding and the budding of microconjugants for sexual reproduction.

Figure 3.12: Anisogamous Sexual Reproduction in a Colonial Flagellate. t — trophic colony, z — zygote.

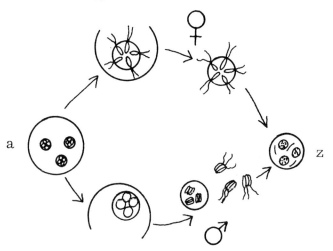

Source: Information from Coleman (1980).

From the foregoing consideration of sexual reproduction it is obvious that sexual competence has a limited distribution among free-living Protozoa, and the majority of species capable of a sexual phase are primitive in having isogametes. Since most protozoans which are capable of sexual reproduction are small, there is no real advantage in having macro- and microgametes. As organisms, particularly multicellular plants and animals, increase in size the chance of a zygote formed by the fusion of microgametes surviving to produce a mature individual decreases. The evolution of anisogamy had been considered by Maynard Smith (1978), who suggests that populations producing microgametes became invaded by macrogamete producers. Once invaded, maximum fitness was achieved by microgametes fusing only with macrogametes. Macrogametes have not evolved the ability to fuse with other macrogametes, even though this would enhance fitness, because the reduced motility of macrogametes would reduce the chances of finding a suitable gamete to fuse with and there would be an increased risk of inbreeding. Within the Volvocida we are able to observe the progression from isogamy to anisogamy in relation to colony size. Small colonial species are isogamous, medium-sized species are slightly anisogamous but with no disassortative fusion, while large species are anisogamous and practise disassortative fusion. However, Maynard Smith (1978) makes the point that further work is required to test this model.

The ability to undergo sexual phase in the life-cycle, even the simplest form of isogamous reproduction, confers considerable genetical advantage on a population, because the chances of successful mutation is much enhanced during the genetic processes occurring in the cell during sexual reproduction.

D. Respiration

Microorganisms, such as Protozoa, are able to obtain the oxygen they require for metabolism from the surrounding medium by diffusion. No special structures or pigments for obtaining and transporting oxygen are necessary. The dependence on diffusion for the entry of oxygen into the cell, however, limits the size a protozoan cell can achieve and still function successfully. Krogh (1941) suggested that when metabolism is high, diffusion alone can provide sufficient oxygen from a normal atmosphere for animals up to 1 mm in diameter. A sphere, however, is an unsatisfactory shape for an organism dependent solely on diffusion. Most of the large species of Protozoa with a fixed shape are elongated, thereby presenting a large surface area to the medium and thus improving conditions for diffusion exchange. Large, naked amoebae not only possess a large surface area, but are also able to bring various portions of the cytoplasm near the surface during the protoplasmic streaming involved in locomotion. Most species are probably obligatory aerobes, though ecological evidence suggests that some may be facultative anaerobes, since many survive prolonged anoxia in an active state in the wild. Recent research has shown, however, that some of the species commonly found in conditions of low oxygen availability, and previously thought to be facultative anaerobes, are in fact obligatory anaerobes lacking cytochrome oxidase or mitochondria with cristae or tubuli (Fenchel et al., 1977).

Body size, or in the case of Protozoa cell size, is an important endogenous factor affecting the respiration of animals. The relationship can be demonstrated for a variety of invertebrates and vertebrates by relating the logarithm of metabolic rate to the logarithm of body weight in a linear regression. The regression coefficient or slope (b), which is usually around 0.75, expresses the relationship between metabolism and weight. The b value may vary within a species in response to environmental factors such as temperature, and biological factors such as nutritional status (i.e. starved or well-fed). It follows that if respiratory data for an organism are to be significant and of value they must be related to the size or the weight of that organism. Respiratory studies on Protozoa have suffered in this respect because of technical difficulties. Until the récent widespread use of cartesian diver microrespirometry, the respiration rates of protozoan species were determined in conventional respirometers, usually of the Warburg type, which necessitated the use of a 'protozoan soup' containing thousands of individuals. Not only was it impossible to make any valid correlation between respiration rate and organism size, but other problems such as bacterial contamination, the mechanical effects of overcrowding, the build-up of metabolites and lack of means of allowing for protozoan reproduction during an experiment, all contrived to introduce considerable experimental error. The net result is that, although there are many data on protozoan respiration in the literature, the majority are fraught with the problems outlined above.

Microrespirometry techniques, such as the cartesian diver, are now becoming more widely employed by protozoologists and allow the respiration rate of an individual or small groups of Protozoa to be determined and related to cell size or weight. Furthermore during cartesian diver microrespirometry, the animals are observable through a microscope so that any reproductive or locomotory activity can be quantified.

(i) Aerobic Respiration

Biochemistry. Figure 3.13 shows a simplified scheme of the metabolic pathways of carbohydrate metabolism in Protozoa (Ryley, 1967). Many of the transformations are really multistage processes, some of which are reversible and some involving hydrogen and phosphate transfers which are not indicated in Figure 3.13. The process of glycolysis, that is the pathway from glucose to pyruvic acid, involves the production of two energy-rich bonds in the form of ATP, and the reduction of two molecules of nicotinamide-adenine dinucleotide (NAD). Polysaccharide and phosphorylated glucose are interconvertible by means of phosphorylase and uridine diphospho-glucose (UDPG). Reduced NAD must be reoxidised in order to maintain glycolytic activity, which in aerobic respiration is achieved by energy-yielding electron transfers along the respiratory chain by a complex of flavoprotein and iron-containing enzymes. Under anaerobic conditions, however, further transformations of pyruvic acid or dihydroxyacetone phosphate take place. The latter may be converted to glycerol, while pyruvic acid may be reduced to lactic acid or ethanol and carbon dioxide. Carboxylation reactions at the level of pyruvic acid also often take place, followed by reduction and the eventual formation of succinic acid.

During aerobic metabolism, pyruvic acid may be oxidised and decarboxylated to acetyl CoA, which is a metabolically active form of acetic acid. Acetyl CoA may be converted to free acetic acid or alternatively may be completely oxidised by the tricarboxylic acid cycle to carbon dioxide and water. The tricarboxylic acid cycle is energy producing and involves transformations between nine di- or tricarboxylic acids. Four of the reactions are oxidative steps and two involve carbon dioxide liberation.

The glyoxylate cycle involves four of the reactions of the tricarboxylic acid cycle in addition to two specific enzymes. During the cycle two molecules of 'active' acetate are converted to one molecule of succinate. Further transformations of succinate to a variety of substances by cycle enzymes may occur. The substances thus formed are important in synthetic reactions.

The biochemical study of metabolism in Protozoa is hampered by the need for large quantities of pure material for experimental analysis. The mass culture of Protozoa under axenic conditions is ideal for this purpose. However, of the enormous number of free-living species only *Tetrahymena* and *Acanthamoeba castellanii* can be reliably obtained in simple axenic culture. In *Tetrahymena*, which is probably typical of free-living ciliates, energy metabolism under aerobic conditions appears to depend on the utilisation of fat and protein and the presence of extracellular substances. Carbohydrate may be assimilated, but it is apparently not used to any great extent for energy purposes, and under suitable environmental conditions fat may be converted to storage glycogen – glyconeogenesis (Ryley, 1967).

From the evidence available it seems that many types of Protozoa have a typical cytochrome respiratory system. However, one striking feature of carbohydrate metabolism is that in many cases the breakdown is incomplete and results in the formation of acid. The phytoflagellates differ metabolically from other free-living Protozoa in having photosynthesis and often heterotrophy combined in a number of ways.

The genus *Euglena* typifies the bridge between autotrophic and oxidative heterotrophic metabolism. Some species, e.g. *Euglena gracilis* var *bacillaris*, grows at the same rate photosynthetically as it does in the dark by the oxidation of organic carbon sources, whereas other species are obligate autotrophs. Evidence indicates that oxygen consumption by *Euglena* and its relatives occurs via a cytochrome system similar to that found in the majority of aerobic organisms (Danforth, 1967).

Metabolic Rates. The relationship between metabolism and weight in an organism is definable by the simple exponential formula

$$M = kW^b \text{ or } \log M = \log k + b \log W$$

where M is oxygen uptake per unit time, W is body weight, k and b are constants. In a group of animals for which k and b have common values, a plot of log M against log W will yield a straight line for which the slope is defined by the exponent b and the intercept of the line on the ordinate is at log k. When the regression coefficient b is 1.0, metabolism is simply proportional to weight. However, this is rare and b is usually less than 1.0. Zeuthen (1953) reviewed a wide range of data on the relationship between weight and metabolism and concluded that a value of b = 0.70 was characteristic of unicellular organisms, i.e. bacteria, Protozoa and the eggs of some marine metazoa. Hemmingsen (1960), however, concluded from his critical review of the available information that with the exception of a possible short transitional range of 0.1 μg to 1.0 mg where b may approach 1.0, a value of 0.75 is universal among animals. A more recent consideration of the literature

Figure 3.13: Metabolic Pathways of Carbohydrate Metabolism in the Protozoa.▷ Key to enzymes used in pathways: 1, phosphorylase — degradation and perhaps synthesis of polysaccharide; 2, UDP synthesis of polysaccharide; 3, phosphoglucomutase; 4, hexokinase — uses ATP; 5, phosphofructokinase — uses ATP; 6, aldolase; 7, α-glycerophosphate oxidase — uses molecular oxygen; 8, α-glycerophosphate dehydrogenase — uses NAD; 9, glucose-6-phosphate dehydrogenase — reduces NADP; 10, Entner-Doudoroff cleavage; 11, glyceraldehyde-3-phosphate dehydrogenase — reduces NAD; 12, pyruvic oxidase; 13, oxalacetic decarboxylase; 14, lactic dehydrogenase — uses NADH; 15, phosphoenolpyruvic carboxy-kinase — requires inosine diphosphate and fixes carbon dioxide; 16, malic enzyme — uses NADPH and fixes carbon dioxide; 17, aconitase; 18, isocitric dehydrogenase — uses NADP; 19, succinic dehydrogenase — no coenzyme involved; 20, fumarase; 21, malic dehydrogenase — uses NAD; 22, isocitratase — liberates succinate; 23, malate synthase — uses acetyl CoA.

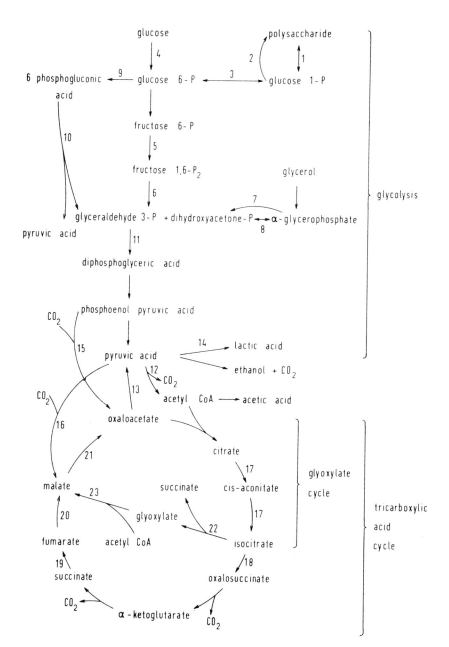

Source: Ryley (1967), with the permission of Academic Press, New York.

by Phillipson (1981) has produced a value of b = 0.66 as being typical of unicellular ectotherms.

Metabolism, however, is not simply proportional to weight or size. This is evident from the fact that with increasing size in a given species, under a constant set of conditions, oxygen consumption declines on a per unit weight basis. If metabolism was proportional to weight, then oxygen uptake per unit weight per unit time should be constant for similar animals of different sizes. Other factors are involved and there may be a tendency for metabolism to be a reflection of surface area rather than body weight in some animals.

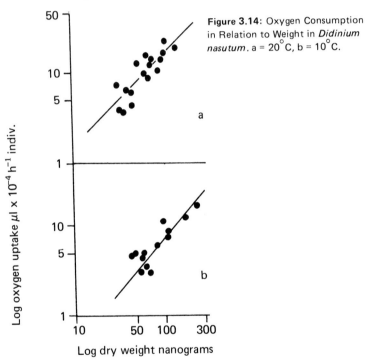

Figure 3.14: Oxygen Consumption in Relation to Weight in *Didinium nasutum*. a = 20°C, b = 10°C.

On an intraspecific basis, attempts to demonstrate a relationship between metabolism and weight in ciliated Protozoa have shown that the typical linear relationship characteristic of Metazoa exists in protozoans (see Figure 3.14). The values for b in ciliates, shown in Table 3.1, vary considerably from 1.0 for *Didinium nasutum* at 20°C to 0.60 for *Stentor coeruleus* at 15°C. Some variation may be attributable to the technique adopted and this was dictated by the sizes of the ciliates involved. *Tracheloraphis* sp. is a small species and 9-11 individuals were contained in each cartesian diver, so that both metabolic rates and weights are average values for small groups of ciliates. The same is true for *Didinium* where 2-7 individuals per diver were investigated. The data for *Stentor*, however, are based on individual metabolism and weight determinations because this is a very large ciliate. The accuracy of the cartesian diver technique and its sensitivity (about $1 \times 10^{-4} \mu l\ O_2$) probably counterbalances the inaccuracies of mean oxygen uptake and weight values, especially since care was exercised in selecting similar-sized individuals for each group of ciliates investigated. The b values in Table 3.1 probably reflect the relationship between weight and metabolism for these species.

Table 3.1: b Values for Ciliated Protozoa Derived from Microrespiration Studies Using the Cartesian Diver Microrespirometer

Species	Temperature	b	Author
Stentor coeruleus	20°C	0.67	Laybourn (1975b)
''	15°C	0.60	''
Didinium nasutum	20°C	1.00	Laybourn (1977)
''	15°C	0.98	''
''	10°C	0.96	''
Tracheloraphis sp.	20°C	0.74	Vernberg & Coull (1974)

The differences between the b values listed for ciliates in Table 3.1 are probably a function of their activity in relation to size. For example, the metabolic rate of *Stentor coeruleus* is very low when compared with that of *Didinium nasutum* of similar size. A small *Stentor* weighing about 200 ng dried weight has a respiration rate of $0.92 \times 10^{-4} \mu l\ O_2\ h^{-1}$ at 20°C, whereas a large *Didinium* of the same weight has a rate of oxygen uptake in the region of $27.5 \times 10^{-4} \mu l\ O_2\ h^{-1}$ at 20°C. *Stentor* is a large trumpet-shaped slow-moving ciliate, which can be up to 1 mm long, and spends a great deal of its time in a sedentary position attached by a holdfast to a substrate, whereas *Didinium* is an extremely active swimming protozoan predator, feeding mainly on *Paramecium* sp. The metabolic rate of *Didinium* is exceptionally high for a protozoan and unusually here b is 1.0 or nearly 1.0 depending on temperatures.

Interspecific relationships between size and metabolism have been investigated by Sarojini and Nagabhusham (1967). These authors studied 13 species of ciliate and were unable to establish any linear trend between weight and respiration rate. Later similar work by

Laybourn and Finlay (1976) considered five species of ciliate and incorporated data on other species (Laybourn, 1975b, 1976c) at a range of temperatures similar to those encountered in the natural environment. Figure 3.15 shows a very clear linear relationship between weight and metabolism on an interspecific basis. Furthermore the relationship holds over a range of temperature from 8.5°C to 20°C. The latter study employed cartesian diver microrespirometry, while Sarajini and Nagabhusham used the Warburg respirometer with all its attendant problems for protozoological studies. The lack of any trends in their data was probably a function of the technique used. An interspecific relationship between oxygen consumption and weight has also been demonstrated in naked amoebae (Laybourn-Parry *et al.*, 1980) from data on *Vannella* sp., *Saccamoeba limax, Chaos chaos* (Holter and Zeuthen, 1947) and *Amoeba proteus* (Rogerson, 1978), and more recently including these data and new data on other species (Baldock *et al.*, 1982).

Figure 3.15: The Relationship between Respiration Rate and Cell Size in Seven Species of Ciliate: *Tetrahymena pyriformis, Vorticella microstoma, Paramecium aurelia, Spirostomum teres, Fontonia leucas, Stentor coeruleus* and *Podophrya fixa*. a = 20°C, b = 15°C and c = 8.5°C.

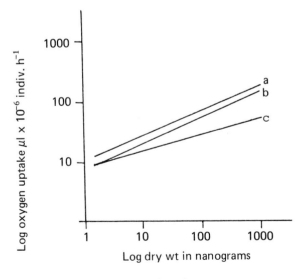

Each species appears to have a distinct temperature range, within which there is an optimum temperature for metabolism. Moreover, the temperature tolerances of species differ, so that some have a wide temperature range, some are thermophilic and others are cold temperature species. Throughout the course of a year in climates where seasonal temperature changes occur, any particular niche will be occupied by a succession of species each succeeding the other as their optima are exceeded. Table 3.2 shows metabolic rates related to temperature for a range of ciliate and amoeba species. It is clear that while some species such as *Paramecium aurelia* and *Tetrahymena pyriformis* are warm temperature species with rates of oxygen consumption increasing in relation to increased temperature, other species, e.g. *Vorticella microstoma* and *Spirostomum teres*, have an optimum temperature for metabolism near $15°C$.

Comparison of the oxygen consumption rates in various taxonomic groups within the Protozoa shows considerable variability, as inspection of the values related to cell size (weight) in Table 3.2 shows. Amoebae generally appear to have higher respiratory rates than ciliates of the same weight. Given the very different physiological and taxonomic characteristics of amoebae and ciliates it is perhaps not surprising that such disparities in metabolic rate exist. Such differences may be a function of higher locomotory and food acquisition costs. However, even within the ciliates the reported values for a single species may vary considerably. *Tetrahymena pyriformis*, for example, has rates varying from 8.7-21.1 x 10^{-6} μl h^{-1} per individual at 8.5-20.0°C (Laybourn and Finlay, 1976), through 60-120 x 10^{-6} μl h^{-1} per individual at 30°C (Finlay *et al.*, 1983) to 120-780 x 10^{-6} μl h^{-1} per individual at 26°C (Lovlie, 1963). Very often the differences result from temperature, species strain used, nutritional status or the technique employed and sometimes the factor of bacterial contamination may be involved.

Q_{10} values (Table 3.2) are useful in characterising the magnitude of increase in metabolic rate as temperature rises. It has been suggested (Wieser, 1973) that low values for Q_{10} are characteristic of the optimum part of an organism's temperature range, whereas large Q_{10} values are suggestive of temperature ranges towards the limits of the metabolic functioning of an animal. Accepting this view, which seems physiologically sound, the values shown in Table 3.2 give some indication of the temperature tolerances of each species.

Temperature is obviously a fundamental factor in controlling the metabolic rates of Protozoa, as it is in other ectothermic organisms, but other factors may also play a role in determining the level of metabolic

Table 3.2: Oxygen Consumption in Ciliates and Amoebae Related to Cell Weight and Temperature.

Species	Temp. °C	Resp. μl x 10^{-6} h^{-1} indiv.	Dry wt. ng	Q_{10}	Authors
Tetrahymena	8.5	8.7	1.64	1.45	Laybourn &
pyriformis	15.0	11.0	1.38	3.39	Finlay (1976)
	20.0	21.1	1.36		,,
Vorticella	8.5	12.2	4.58	1.52	,,
microstoma ʔ	15.0	16.6	5.12		,,
	20.0	9.8	4.43	—	,,
Paramecium	8.5	24.3	22.38	1.01	,,
aurelia	15.0	26.6	32.84	1.70	,,
	20.0	33.1	26.86		,,
Spirostomum	8.5	11.2	36.18	12.98	,,
teres	15.0	59.8	35.55		,,
	20.0	41.2	59.29	—	,,
Frontonia	8.5	34.0	84.36	2.24	,,
leucas	15.0	57.5	106.60		,,
	20.0	50.9	104.37	—	,,
Stentor	15.0	150.0	860.00	3.82	Laybourn
coeruleus	20.0	350.0	1,100.00		(1975b)
Podophrya	15.0	8.7	8.14	5.76	Laybourn
fixa	20.0	21.1	8.42	3.72	(1976b)
	25.0	40.9	10.39		,,
Strombidium spp.	22.0	128.0	1.50	—	Klekowski &
Tiarina fusus	24.0	194.0	27.00	—	Tumantseva
Diophrys spp.	23.0	163.0	21.00	—	(1981)
Saccamoeba	10.0	425.0	0.82	0.12	Laybourn-
limax	15.0	464.0	1.54	1.22	Parry *et al.*
	20.0	512.0	0.68	1.33	(1980)
	25.0	590.0	1.04		,,
Vannella spp.	10.0	968.0	1.29	7.36	,,
	15.0	728.0	1.17	2.01	,,
	20.0	510.0	1.73	1.77	,,
	25.0	188.0	1.18		,,
Amoeba	10.0	952.0	372.00	—	Rogerson
proteus	15.0	761.0	278.00	—	(1981)
	20.0	1,080.0	178.00		,,

activity. Nutritional status has been demonstrated as affecting meta-bolism in the suctorian *Podophrya fixa* where oxygen consumption per unit body weight decreased during starvation over a 96-hour period (Laybourn, 1976c). In this instance not only did cell volume decrease, but also the metabolic rate per unit cell weight declined. In *Chaos chaos*, on the other hand, the ratio between oxygen consumed and reduced weight was constant and independent of the shrinkage of the amoeba, which usually dies about one month after the onset of starva-tion (Holter and Zeuthen, 1947). This suggests that in *Chaos* the respiratory rate per unit body weight was constant throughout starva-tion. In the wild, however, a mobile species like *Chaos* can search for food when starved, whereas *Podophrya* is by virtue of a sedentary habit dependent on prey items making contact with its feeding tentacles, so that reduced metabolic rate per unit cell weight as starvation progresses may be an adaptation to conserve energy and thus withstand long periods without food.

Flagellates also show marked changes in metabolic rate as their stage of growth, and presumably nutritional status, vary. *Astasia klebsii* in the log phase of growth in culture had a maximum rate of oxygen uptake of 10.25 μl h^{-1} per million cells and a minimum of 4.0 μl h^{-1} per million cells. During the stationary culture phase maximum respiration rate dropped to 4.0 μl h^{-1} per million cells and minimum rates to 2.25 μl h^{-1} per million cells (von Dach, 1942). A recent bioenergetics study on heterotrophic flagellates found respiration rates of 5.0 x 10^{-9} μl h^{-1} per individual in *Ochromonas* and 1.9 x 10^{-9} μl h^{-1} per cell in *Pleuromonas jaculans* at 20°C (Fenchel, 1982). The impact of reduced food concentration is more pronounced in these two species when compared with *Astasia*. In both starved *Ochromonas* and *Pleuromonas* respiration rate dropped to 2-4 per cent of the rate characteristic of the exponential phase of growth.

Population density may affect metabolic rate under laboratory conditions. Pace and Kimura (1944), while investigating respiration in *Paramecium aurelia* and *P. caudatum*, noticed that the rate of oxygen uptake per individual was greater when fewer animals were present. The impact of population density on respiration was considered in more detail by Pace and Lyman (1947), who found that individual oxygen consumption was inversely proportional to population density in *Tetrahymena geleii*. In the wild, protozoan populations probably rarely reach the densities considered in such laboratory studies, but there may be the odd unusual occurrence when enriched conditions stimulate high population numbers causing a depression in individual metabolic rates as a result of mechanical stimuli or an accumulation of metabolites acting as inhibitory factors.

(ii) Anaerobic Respiration

Obligatory anaerobic metabolism is known among some parasitic Protozoa, but the condition is not widely documented in free-living species. Certainly Protozoa have repeatedly been reported from conditions where oxygen is limited or not available. Strict anaerobiosis is rare in eukaryotes and consequently it has been assumed, probably correctly, that the majority of species found in these circumstances were facultative anaerobes surviving difficult conditions by reverting from an aerobic metabolism to less efficient anaerobic metabolism.

One particular ecological group of ciliates associated with anaerobic sulphide-containing sediments were recently examined for cytochrome oxidase activity. The species included members of the genera *Sonderia*, *Metopus, Plagiopyla, Parablepharisma* and *Caenomorpha*. All were found to be lacking in cytochrome oxidase activity and in addition such species did not possess typical protozoan mitochondria (Fenchel *et al.*, 1977). Although most of these so-called 'sulphide ciliates' lack mitochondria, they do contain microbodies. These are oblong, spherical or irregularly-shaped membrane-bound organelles about 1-2 μm in length. Most of these anaerobic ciliates also harbour ecto- and endo-symbiotic bacteria which may possibly utilise the metabolic end-products of the ciliate metabolism for growth and energy-yielding processes. Fenchel *et al.* (1977) suggest that these ciliates, which incidentally die on exposure to oxygen, have evolved from aerobic forms. The anaerobic state is primitive, since it is almost certain that the biosphere was once anoxic. The sulphide ciliates have reverted during evolution to an anaerobic metabolism in order to exploit an ecological niche inhabited in marine sediments primarily by prokaryotes. Other species of ciliate are frequently found feeding on sulphur bacteria in the vicinity of anaerobic zones, but are in fact aerobes.

E. Osmoregulation and Excretion

Osmoregulation is an essential function in freshwater Protozoa because the osmotic pressure of the cell is greater than the aquatic environment in which these organisms live. There is a net gain of water from the surrounding medium, which if unchecked would result in the eventual swelling and rupturing of the cell. Contractile vacuoles are the organelles which perform osmoregulation and they may be regarded as being analagous to the kidney of higher organisms. With the exception of dinoflagellates, all freshwater Protozoa possess one or more contractile vacuoles. Since marine species do not normally need to osmoregulate, because they are usually in isotonic equilibrium with their environment, marine flagellates and amoebae lack contractile vacuoles. The organelle is present, however, in the majority of marine ciliates.

Kitching (1967) has pointed out that it may be erroneous to assume that marine species do not need to osmoregulate. There is evidence that many marine species are permeable to certain ions in sea water, including Na^+ and Cl^-. Protozoa have a high proportion of potassium to sodium and a low internal concentration of sodium is probably maintained at a low level by the 'sodium pump'. The potential of the sodium pump as a means of lowering internal osmotic pressure depends on the concentration of sodium in the external medium. Thus the potential of the sodium pump is considerable in a marine environment but small in fresh water. Kitching, therefore suggests that the contracting vacuole is unnecessary in marine Protozoa, and its presence in some marine species is suggestive of a freshwater ancestry.

Water is not only gained by osmosis through the cell membrane, but is also taken into the cell in food vacuoles and formed metabolically during respiration within the cell. Water taken in during food vacuole formation probably does not greatly affect the output of the contractile vacuole. Although food vacuoles shrink initially, by the time their undigested contents are expelled they have resumed their original size, or nearly so. An equivalent amount of water to that taken in during food vacuole formation is expelled with undigested material. Kitching (1956) believes that metabolic water can be calculated on the assumption that each mole of oxygen consumed gives rise to one mole of water as in the oxidation of glucose. Most of the water expelled from a protozoan cell is water gained by osmosis and this can be demonstrated experimentally. In many ciliates an increase in osmotic pressure causes an increase in the rate of vacuolar output. There is experimental evidence to suggest that salts enter and leave the cell, and that over a period of time a new steady state is set up when a protozoan is subjected to osmotic stress.

The structure of the contractile vacuole varies among the protozoan groups, but in almost all cases the vacuoles are surrounded by a system of fine membranous tubules and vesicles which is termed the spongiome. In the ciliates the most complex arrangement of the spongiome has evolved and the contractile vacuole discharges through a permanent pore which is supported by two sets of microtubules. Patterson (1980) has defined four structural types of contractile vacuole occurring within various protozoan groups. The arrangements illustrated in Figure 3.16a and b are found in the ciliates. The Kinetofragminophoran ciliates (Figure 3.16b) show a less well-developed spongiome than the Oligohymenophorans (Figure 3.16a), where the microtubules of the spongiome feed collecting canals which discharge into the contractile vacuole. Part of the spongiome is distensible and appears as ampullae before the expulsion of vacuolar liquid (Patterson, 1976, 1977).

In large amoebae (Figure 3.16c) the spongiome forms a discrete layer around the contractile vacuole and is made up of a large number of small vesicles (20-50 nm). Surrounding this area many amoebae have a zone rich in mitochondria, presumably supplying energy for the osmoregulatory processes. Unlike the arrangement in ciliates the membrane of the contractile vacuole breaks up at systole to form vesicles which expand and later coalesce to reform the contractile vacuole (McKanna, 1973). Thus membrane is essentially recycled. Small amoebae and flagellates (Figure 3.16d) also have a spongiome composed of a layer of irregular vesicles and tubules, which enlarge and fuse to form the contractile vacuole, which breaks up at systole to reform later (Patterson, 1980). The contractile vacuole structure of the Protozoa thus shows a rather striking progression from a relatively simple system (Figures 3.16c, d) as found in amoebae and flagellates to the establishment of a permanent contractile vacuole with a discharge pore as evolved by the ciliates. Within the ciliates a progressive increase in complexity is also apparent, from the arrangement in the kinetofragminophorans where collecting canals are not present to the elaborate system of collecting canals fed by microtubules found in oligohymenophoran ciliates.

The contractile vacuole cycle is characterised by three stages: the diastole or filling process, the systole (which is a phase of vacuolar contraction) and finally the expulsion of the vacuolar fluid. The end of diastole and the beginning of systole is marked by the rounding up of the contractile vacuole, that is the point at which an irregular form adopts a regular shape (Patterson and Sleigh, 1976). In the advanced ciliates the ampullae are apparent around the contractile vacuole after systole is initiated and their appearance is associated with a slight decrease in vacuolar volume attributable to some backflow of fluid from the contractile vacuole into the spongiome, caused by contractile elements around the vacuole. The ampullae persist after expulsion of the vacuolar fluid has occurred, after which they pass their contents

Figure 3.16: Various Types of Contractile Vacuole Found in the Protozoa.
a: The contractile vacuole of the ciliates.   b: The system found in kinetofragminophoran ciliates.   c: The arrangement typical of large amoebae.
d: The arrangement found in flagellates and small amoebae. am — ampulla,
cc — collecting canal, m — mitochondria, mb — microtubular bands,
pm — discharge pore.

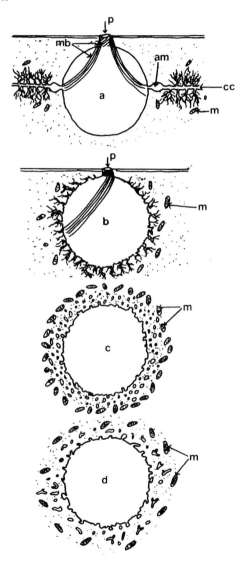

Source: Based on Patterson (1980).

into the vacuolar fluid has occurred, after which they pass their contents into the vacuole at diastole. This however, does not occur until the contractile elements around the contractile vacuole have relaxed, so there is some delay between expulsion and the passage of fluid from the ampullae to the vacuole. There are some variations on this basic pattern due to morphological variation among the ciliate groups (Patterson, 1980).

In the amoebae, flagellates and most of the kinetofragminophorans the vacuole develops by the fusion of smaller contributory vesicles. In some cases these contributory vesicles arise in the mass of cytoplasm occupied by the contractile vacuole before expulsion; alternatively the vacuole reappears but its means of refill is not apparent so that the contributing vesicles must be very small. Some work on kinetofragminophorans suggests that the contractile vacuole may not fragment at expulsion, and that the apparent vesicles that refill the vacuole are the result of uneven filling of the collapsed vacuole (Patterson, 1980).

The exact mechanism which effects contractility and the consequent expulsion of vacuolar contents is not understood, nor do we know how water is collected from the cytoplasm. The widely-accepted view is that fluid is collected at cytoplasmic sites, independent of the contractile vacuole, and transported to the vacuole. Furthermore there would seem to be a continuous gain and loss of membrane from the vacuolar complex. Recently Patterson (1980) has proposed a new concept of the contractile vacuole based on the evidence gained from ultrastructural studies on the vacuolar and spongiome structure. He suggests that the contractile vacuole complex is a discrete organelle, which is stable and has no massive continuous recruitment or loss of membrane from its complex. Furthermore the evidence points to the elements in the complex as the sites for the production of fluid destined for the contractile vacuole.

The role of vacuolar output varies in relation to the size of the protozoan cell and temperature. Kitching (1956) argues that increased vacuolar output in response to increasing temperature is the result of changes in the permeability of the cell membrane to which the contractile vacuole quickly responds. Table 3.3 shows the rate of vacuolar output in a number of species. *Amoeba proteus* is a large species of amoeba with a mean cell volume in the region of $1,000 \times 10^3 \ \mu m^3$-$1,400 \times 10^3 \ \mu m^3$ at 20°C and consequently takes a long period, up to 13 hours, to expel the equivalent of its own body volume, whereas a smaller species like *Paramecium caudatum* with a cell volume of approximately $300\text{-}400 \times 10^3 \ \mu m^3$ takes 49 minutes to achieve the same.

Table 3.3: Contractile Vacuole Output in Some Protozoa

Species	Temperature °C	Rate of Output $\mu m^3 sec^{-1}$	Time to Eliminate Equivalent of Body Volume
Amoeba proteus	19-27	54-109	3.9-13.2 hrs
Paramecium caudatum	15-23	54-258	15-49 mins
Carchesium aselli	14.5-16	6-20	25 mins

Source: Data from Kitching (1956).

In relative terms small protozoans have higher rates of osmoregulation than larger species. The amount of water which must be expelled is a function of the water entering the cell by osmosis and in small protozoans, where the surface area in relation to cell volume is large, relatively greater volumes of water enter the cell when compared to large protozoans where the surface area to volume ratio is lower.

4 MOVEMENT

A. Introduction

Most free-living Protozoa possess the ability to perform locomotion, and even the sedentary forms are capable of movement, though in the Suctoria active locomotion from one locality to another is restricted to the early stages of the life-cycle. Different means of achieving movement have evolved among the free-living groups, the most marked contrast being between the Sarcodina on the one hand and the ciliated and flagellated Protozoa on the other. Characteristically the sarcodines possess pseudopodia, which vary structurally among the classes and orders of naked and shelled amoebae. Inevitably the gross movements between amoebae with skeletal structures and those without will appear distinctly different, since skeletal structures impose limitations on the type and speed of movement, as a function of test or shell shape and weight. The ciliates and flagellates have locomotory organelles which are structurally similiar. Typically flagellates have only one or two flagella, whereas in the ciliated Protozoa cilia are numerous and consequently a complex system co-ordinating ciliary beat has evolved.

Movement serves a variety of functions. The majority of organisms require to move around their environment to search for food, or towards favourable chemical and physical conditions as well as away from unfavourable stimuli. Movement is necessary in the process of feeding, involving special movements of locomotory appendages intended to capture or collect items of food and direct them into the cell. In all Protozoa locomotion is achieved by propelling the cell through the environment by a variety of locomotory processes; during feeding in many ciliates and flagellates, however, the aqueous environment is propelled over the cell, or onto a particular area of the cell surface. Special movements are also frequently involved in the sexual phases of the life-cycles of those Protozoa possessing sexual competence.

The first observations on locomotion in Protozoa were made by

Antony van Leeuwenhoek, who even observed the movement of cilia. In describing a ciliate he wrote, 'Their belly was flat provided with divers incredibly thin little feet or little legs, which ever moved very nimbly, and which I was able to discover only after sundry great efforts, and wherein they brought off incredibly quick motions' (from a translation by Dobell, 1960). Until the advent of the electron microscope and subsequent suitable fixation and staining techniques, in conjunction with the development of sophisticated biochemical analytical procedures, the structure of the organelles and the mechanisms involved in the movement of ciliates, amoebae and flagellates were shrouded in mystery.

We are now aware that one of the major suborganellar structures involved in support and elongation of locomotory organelles in protozoans is the microtubules of the cortex. Lynn (1981) suggests that the evolution of microtubules was undoubtedly one of the most important events for eukaryote protists, because it allowed form to be maintained and modified. The applications of microtubules in cellular processes are wide and in locomotory function they have reached their most complex organisation in the ordered cortex of the ciliates.

A considerable literature now exists on the organellar and suborganellar structure of locomotory organelles, particularly in ciliates. Since these protozoans are the most complex, they have inevitably been the focus of attention. Our knowledge of the mechanisms of movement, although reasonably comprehensive, is by no means complete. Certain aspects of locomotory physiology and biochemistry are still vague; for example, we have very little information on how patterns are formed in developing organellar complexes.

B. The Structure of Locomotory Organelles

(i) Ciliophora and Mastigophora

The complexity of the ciliate and flagellate cortex has rendered its study a difficult task, in terms of both structure and function. Investigations have focused at two levels, first on structure and wave patterns, and secondly on the chemical properties of structural components and the chemical mechanisms involved in movement. The structure of cilia and flagella is fundamentally similar. These organelles have a circular or elliptical cross-section with usually a constant diameter of 0.15-0.3 μm up to a short distance from the rounded or pointed tip (Holwill,

1966). Within the cilium or flagellum is the axoneme which is made up of two centrally-positioned microtubules, surrounded and joined by crossbridges to nine doublet microtubules (Figure 4.1). Variations do occur in axonemal morphology, but typically these variations are restricted to the axial region occupied by the two central microtubules (Warner, 1974).

Figure 4.1: The Typical 9+2 Arrangement of Tubules in a Flagellum (x 150,000).

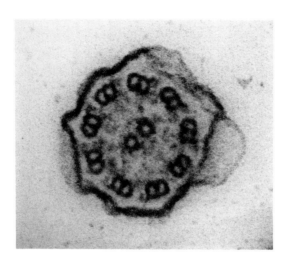

Each of the peripheral doublets consists of an A- and B-subfibre or microtubule. Figure 4.2 shows clearly that while subfibre A is a complete tubule, subfibre B is incomplete and shares a common wall with subfibre A. Attached to each A-subfibre are paired inner and outer ATPase or dynein arms (Figure 4.2). The dynein arms together with 10-degree doublet skewing confer the axoneme with the property of enantiomorphism. Thus if the axoneme is viewed in transverse section and from the base of the organelle to its tip, the enantiomorphic form reveals the dynein arms directed in a clockwise manner and the doublet skew, whereas when viewed from tip to base the direction is reversed counter-clockwise (Satir, 1965; Warner, 1974).

The central doublet of the axoneme is surrounded by a sheath from which radiate spokes or radial arms, each one attaching to the A-subfibre of each peripheral doublet. Other crossbridges join adjacent doublets peripherally. The radial spokes vary in number depending on

Figure 4.2: Diagrammatic Representation of a Portion of the Tubules in a Cilium or Flagellum. CM — central microtubule, I A — inner arm, INL — inter-doublet link, LH — link head, OA — outer arm, RL — radial link, SA — subfibre A, SB — subfibre B.

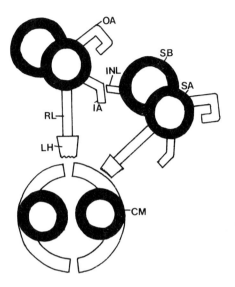

Source: Based on Warner (1974).

species, occurring in pairs in *Chlamydomonas* (Witman *et al.*, 1978) and in triplets in *Tetrahymena* (Warner and Satir, 1974).

Apically the peripheral fibrils in the axoneme extend further than the central doublet, and near the tip the fibrillar arrangement loses its ordered nature. At the base of the flagellum the two central fibres of the axoneme end at a granule or axosome near a basal plate, while the nine peripheral doublets pass through the plate into the basal body. Here they take the form of triplets and twist relative to the doublet formation (Figure 4.3). The bases of the flagella bear fibrous structures which extend into the cytoplasm. A network of fine fibres forms a root system attaching to structures such as the nucleus, plastids or cell membranes. In some cases coarse fibres or ribbon-like structures link the flagellar base to other regions of the cell (Holwill, 1966).

In the Ciliophora the kinetid is the fundamental component of the cortex from which the cilium arises. The basic component of the kinetid is the kinetosome or basal body. The distal ends of the kinetosome microtubules give rise to the peripheral doublet microtubules of the cilium axoneme (Allen, 1969). The kinetosome is composed of

Figure 4.3: Longitudinal Section of a Flagellum. bb — basal body, bp — basal plate, cm — cell membrane, pd — peripheral doublet, cp — central pair of microtubules, cyl — cylindrical cartwheel structure.

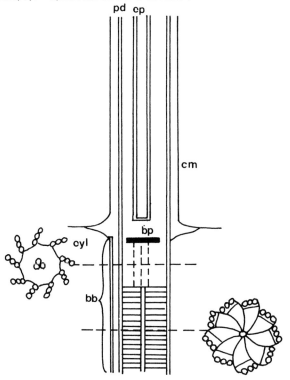

Source: Based on Holwill (1966).

nine microtubule triplets (Figure 4.4), the same arrangement as in the basal body of the flagellum. The kinetosome is capped by a terminal plate, the appearance of which differs among ciliate species, being very dense in some species and less dense in others (Lynn, 1981). The terminal plate is separated from the axosomal plate by a transitional zone. Above the axosomal plate lies the axosomal granule (Figure 4.4). This granule or axosome is large in most ciliates and one of the central pair of axonemal tubules rests on it. The axosome of flagellates is much reduced compared to the axosome of most ciliates (Hibberd, 1979). Distal to the axosomal plate and the axosome, lies the axosomal shaft with its almost universal characteristic 9 + 2 arrangement of microtubules or fibres.

Figure 4.4: Longitudinal Section of a Ciliate Kinetid with Transverse Sections at Various Levels. al — pellicular alveoli, ax — axosome, axp — axosomal plate, cm — cell membrane, cw — cartwheel structure, cp — central pair of microtubules, ep — dense epiplasm, kd — kinetodesmal fibril, pd — peripheral doublet of microtubules, ps — parasomal sac, pc — post-ciliary ribbon of microtubules, sp — secondary plate, tp — terminal plate.

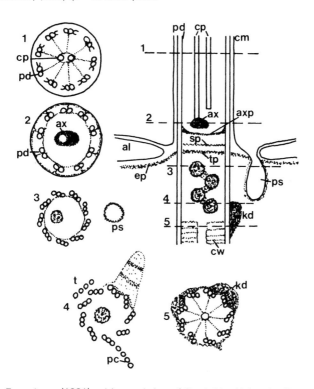

Source: From Lynn (1981), with permission of Cambridge University Press.

Each kinetid has a kinetodesmal fibril, which is essentially a ciliary rootlet. The direction of the kinetodesmal fibril and its form varies among ciliates, but it always originates from the proximal, anterior to lateral right portion of the kinetosome (Lynn, 1975). The kinetodesmal fibril extends parallel to or towards the cell surface. Here ciliates and flagellates differ, because in flagellates the root goes deep into the cytoplasm (Figure 4.3).

Arising from near the proximal, posterior right quadrant of the kinetosome is the post-ciliary ribbon, which extends posteriorly towards the cell surface (Figure 4.4). The ribbon is frequently anchored in dense material at the kinetosome base. This dense material may extend along the ribbon for some distance, as it does in some heterotrichs (Grain, 1968) and in the colpodid *Bresslaua* (Lynn, 1979). Another ribbon of microtubules, termed the transverse ribbon, extends from the kinetosomes of one ciliary row, or kinety, transversely towards the kinetosomes of the kinety on its left. There appears to be considerable variation in the orientation of the transverse ribbon among different ciliate groups. The structure was first described in tetrahymenids, where the ribbon is oriented transversely (Lynn, 1981) – hence the terminology. Parasomal sacs are membranous invaginations, which may be variously positioned in relation to the kinetosome depending on species (Figure 4.4).

From the foregoing description of the structure of cilia and flagella it is clear that microtubules are a fundamental structural component of the locomotory organelles of ciliates and flagellates. Inevitably the chemical structure of the microtubules and their associated structures become an area of intensive study, because elucidation of molecular structure would probably pave the way to an understanding of the mechanism involved in the movement of cilia and flagella. Some of the early pioneering work on ciliary tubules was undertaken by Gibbons (1963) and was later extended by him and his co-workers (Renaud *et al.*, 1968) on the cilia of *Tetrahymena*. The protein isolated from the microtubules, which was named tubulin, was found to have an amino acid composition resembling that of actin, one of the important cytoplasmic contractile proteins which occurs especially in muscle cells. However, not long after the characterisation of tubulin, further work on actin made it clear that the two proteins were distinctly different. At the gross level muscle actin and tubulin have some interesting similiarities. Both proteins contain equimolar amounts of nucleotide di- or triphosphates, both are rich in glutamic and aspartic acids and have a high content of hydrophobic amino acid and free-SH groups, both interact with characteristic ATPases and both form polymers. Close examination, however, dispels any doubts as to the distinct identities of these two separate proteins. The molecular weights are dissimilar – tubulin has a minimum molecular weight of 59,000, which is higher than actin – but more obviously the morphology of the native polymer differs. Actin is a double helix of 55 Å globular units, possessing a repeat period of about 700 Å, whereas a protozoan microtubule is a hollow tube of about 220 Å diameter made up of 12-13 parallel protofilaments of 40 Å monomenic subunits, linearly arranged. Nonetheless the number of intriguing parallels between tubulin and actin may hold some evolutionary significance (Stephens, 1974).

ATPase activity was also found in the axonemes of cilia from *Tetra-hymena* by Gibbons (1963, 1965), and was shown to be localised in the arms of the A-subfibres. The enzyme was given the name dynein, (after dyne – a unit of force), because it was believed to be responsible for the mechanochemical transduction of energy required for move-ment. Dynein was found to hydolyse specifically ATP in preference to other nucleotides; its activity required Mg^{2+} and Ca^{2+} and it was inhibited by EDTA (Gibbons, 1966). Further studies on the protein revealed that the dynein of the outer and inner arms, on the A-subfibre, possessed morphological differences and were composed of different polypeptides, having different molecular weights of 600,000 and 5,400,000. The former consists of globular particles about $14 \times 9 \times 9$ nm while the latter heavier dynein consists of rod-like particles of variable length (Gibbons, 1981).

(ii) Sarcodina

The amoebae have received less attention than the flagellates and ciliates, but nevertheless there is now accumulating a body of informa-tion on the structure and mechanism of movement in naked, shelled and testate amoebae. The movement of the majority of sarcodinids involves the flowing of cytoplasm, which is clearly visible under a light microscope, where the 'granules' in the cytoplasm can be seen moving within the organism. The pseudopodium is the basic locomotory organ-elle, but the manner in which different types of pseudopodia bring about gross movement is variable. There is considerable diversity in the form of the pseudopodia among the different groups of Sarcodina.

Among the naked amoebae and the testate forms, lobose pseudo-podia of a variety of types are found. In some species, e.g. *Naegleria*, only one pseudopodium or pseudopodial lobe is extended at any one time. In *Amoeba proteus* (Figure 1.4) numerous tubular pseudopodia are extruded simultaneously. Tubular forms of pseudopodia also arise from the aperture of the test in testate amoebae such as *Arcella* and *Difflugia* (Figure 1.6). Some naked amoebae retain a sac-like form throughout locomotion and no gross pseudopodia are obvious, e.g. *Pelomyxa*. The mechanisms operating within the cell to form pseudopodia and facilitate movement have been a matter of some debate involving several theories, of which the tail contraction or sol⇌gel transformation and the frontal contraction hypothesis, were most widely considered. This aspect of amoeboid movement will be considered further under Section C.

Modern biochemical analysis and electron microscopy have revealed the presence of contractile elements in the cytoplasm of amoebae. The contractile protein actin has been isolated from various sarcodine species including *Acanthamoeba* and *Amoeba proteus*. Actin is a major contractile protein throughout the Animal Kingdom, and its role in muscle contraction needs no discussion here. Muscle actins and cytoplasmic actins differ probably as a result of different evolutionary pressures on their functional mechanism (Pollard, 1981). All actins have the capability of binding myosin reversibly, and in the presence of ATP of activating the myosin Mg ATPase activity. Cytoplasmic actin, as it occurs in amoebae and other cells, has less affinity for myosin than muscle actin. Myosin is the force-generating energy-transducing enzyme in muscle contraction systems. Unlike the actins, myosins are very variable, and in non-muscle cells are of different shapes and sizes. In *Acanthamoeba*, for example, multiple myosins occur (Pollard and Korn, 1973; Pollard *et al.*, 1978). The question arises as to whether actin and myosin perform the same function in the cells of Protozoa, as they do in muscle cells. It is tempting to assume that this is so. The arrangement of actin and myosin in muscle cells has perfect geometry. In non-muscle cells filaments of actin and myosin can form loosely-organised networks, which has led Pollard (1981) to suggest that, based on mechanical evidence contractile proteins in non-muscle cells can generate tension and motion, by the same sliding filament mechanism which occurs in muscle.

The heliozoans are spherical protozoans possessing numerous long slender pseudopodia termed axopodia. Each axopod has a central structure, the axial rod or axoneme. Axonemes are made up from a large number of microtubules, which are arranged parallel to the long axis of the axopod (Figure 4.5). The longest microtubules occur in the middle of the axonemes and arise from within the cell from various organelles (Tilney and Porter, 1965). The microtubule patterns of heliozoans show variability, but three broad pattern types have been observed. In the first type the microtubules are arranged in two interlocking spirals and the axonemes originate from the nuclear membrane or are found independently in the cortical cytoplasm as in *Actinophryida* (Roth *et al.*, 1970). In the second pattern the microtubules are ordered in distorted hexagons and equilateral triangles, and all the axonemes arise from the central granule or centroplast. This arrangement is typical of the Centrohelida (Bardele, 1975). The microtubules of the third type of pattern are arranged in elongated hexagons, forming a parquet floor type of array, as seen in *Gymnosphaera albida* and *Hedraiophrys horassei* (Jones, 1975; Bardele, 1977). The number of microtubules per axoneme is not constant and can range from as little as six up to 140, depending on species. In places along each axoneme intertubular links occur (Bardele, 1977).

The foraminiferans have a characteristic pseudopodial arrangement of a network of fine pseudopodial strands which are termed reticulopodia and which run into each other, so that there may be considerable variability in the thickness of the reticulopodial strands (Figure 4.5). Microtubular fibrils support the reticulopodia (McGee-Russell and Allen, 1971). It is possible that the fibrils not only support the protoplasmic strands, but may also be involved in movement. Since most foraminiferans carry a large shell, the pseudopodial network probably also contains contractile elements, which together with the microtubules bring about movement.

Figure 4.5a: The Axopods of Heliozoans. Axonemes of microtubules support the axopods. Each axopodium penetrates into the cell where it is attached to cellular structures such as the nuclei in the medulla of the cell. Dense granules and mitochondria are scattered along the length of each axopodium.

C. Chemical Basis of Movement

The discovery of the molecular make-up of protozoan cells, and in particular those elements of the cell involved in movement, has allowed investigation into the chemical reactions which cumulatively bring

Figure 4.5b: The Reticulopodial Network of the Foraminiferan *Allogromia*.

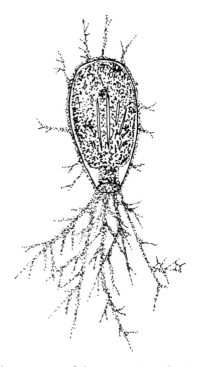

about the overall movement of these organisms. Inevitably the presence of an 'actin-like' protein, the tubulin, in the axonemes of cilia and flagella, and the presence of dynein and its ATPase activity, led researchers to seek parallels, both chemically and morphologically, with the processes involved in muscle contraction. It is now clear, however, that the processes occurring in the axonemes of cilia and flagella constitute a motile system which is different from that operating in the muscle cells of higher organisms. The tubulin–dynein and the actin–myosin systems do have similarities, because both involve the interaction of a distinct high molecular weight ATPase with a hydrophobic nucleotide-containing, globular protein in a polymeric form (Stephens, 1974). The uniqueness of these two systems, both structurally and biochemically, suggests parallel evolution of two contractile systems, one serving at a cellular movement level, the other evolved for locomotion in multicellular organisms.

Actin and myosin have been identified in some species of Sarcodina, and no doubt further investigations will reveal the presence of these proteins in many other amoebae, both shelled and naked. The uniqueness of the actin–myosin reaction would suggest that in amoebae a contractile role is performed by actin, although it by no means resembles the level of force or pattern which occurs in metazoan muscle. What function do such proteins serve in the cytoplasm of single-celled motile organisms, if not a locomotory rule?

(i) Chemical Processes in Cilia and Flagella

The advances in our understanding of the mechanism by which movement is propagated in cilia and flagella have been reviewed by Gibbons (1981). Early theories centred on the assumption that the peripheral doublet tubules possessed the capability of propagating active, localised contractions along their length. The impulses generating the contractions were believed to arise rhythmically at the basal end of one doublet, and to propagate contractile activity to other doublets in the axoneme in a unidirectional manner in cilia, and a bidirectional manner in flagella (Bradfield, 1955). Bradfield suggested that one tubule of each doublet might slide up against another, without either tubule shortening. Around this time also, Huxley and Hanson (1954) proposed a theory of muscle contraction, which became known as the sliding-filament model. This model, which was an exciting idea at the time, clearly attracted Bradfield in his consideration of movement of cilia and flagella.

Investigations using the electron microscope and biochemical procedures in the 1960s and 1970s, provided considerable conclusive evidence for the process of active sliding between flagellar tubules (Satir, 1965, 1968; Summers and Gibbons, 1973). Summers and Gibbons (1973) isolated axonemes from the sperm of sea urchins which they digested briefly with trypsin, in order to disrupt the radial spokes and nexin links in the axonemes. The subsequent addition of ATP brought about the disintegration of the axoneme into separated microtubular doublets, which occurred by extrusion of the tubules from the axoneme by a gradual sliding process. The results of such studies suggested that the dynein arms generate active shearing stress between adjacent doublet tubules. In the normal situation, as opposed to the experimentally isolated axoneme, the shear stresses are resisted and coordinated by the radial spokes and nexin links. In the experimental trypsin-treated axonemes, the nexin links and radial spokes are lacking; hence there is unlimited sliding of the tubules leading to ultimate disintegration of the axoneme. Thus the uniform propagation of waves along the axoneme could be accounted for.

Biochemical studies have shown that the dynein arms rebind to their original sites on the A-subfibre after solubilisation (Gibbons, 1965). Additionally dynein can be demonstrated experimentally to bind onto the B-subfibre, probably to sites with which the dynein arms would normally interact during ciliary or flagellar movement (Takahashi and Tonomura, 1978). Thus a motile role for dynein arms has been shown; they are capable of orientating towards the outer tubule of a doublet, a role which was in any case implicated by virtue of their ATPase activity. Sliding occurs between adjacent doublet microtubules mediated by dynein arms, causing a shearing force exerted towards the tip of the axoneme, resulting in the propagation of a wave along the flagellum or cilium.

There is now considerable evidence to show that internal Ca^{2+} levels are regulated to effect changes in ciliary and flagellar movements. In *Paramecium* the regulatory role of Ca^{2+} was demonstrated by Naitoh and Kaneko (1973). Where the levels of Ca^{2+} are at or below 0.1 μM, the direction of swimming is forwards, but when the levels of Ca^{2+} are above 1 μM a reversal of ciliary beat occurs and the cells move in a backwards direction. These workers were able to demonstrate further the presence of two motile components, one responsible for cyclic beating and activated by $MgATP^{2-}$ and a second responsible for orientating the effective stroke, which was activated by $CaATP^{2-}$.

A similar regulatory function for Ca^{2+} has been revealed in the flagella of *Chlamydomonas* by Bessen *et al.* (1980). They found that isolated axonemes underwent two distinctly different types of movement depending on the concentration of Ca^{2+} in the reactivation solution. In the presence of 1 μM or less free Ca^{2+}, the axonemes propagated nearly planar, asymmetrical waves, which would result in the flagellates swimming in circles of relatively small diameter. The presence of 0.1 μM free Ca^{2+} propagated nearly planar symmetrical waves, which would result in the organism moving through the medium in a straight line. Thus appropriate stimulation of the cell leads to a depolarisation of the cell membrane, which in turn results in the opening of voltage-sensitive Ca^{2+} channels in the cell membrane. Ca^{2+} flows in, leading to a high internal concentration of Ca^{2+} and an alteration of ciliary beat. The result of a high internal concentration of Ca^{2+} is the deactivation of Ca^{2+} channels, and a rapid lowering to a resting level by the action of Ca^{2+} pumps in the cell membrane, causing the flagella to return to their normal beat pattern.

Our understanding of the molecular mechanisms responsible for the sliding of adjacent doublet microtubules is still incomplete. The picture we have suggests an ATP-driven crossbridge cycle for dynein in cilia and flagella, similar, in many respects, to that which is understood to occur in the myosin cross-bridge cycle of muscle in higher organisms. As yet the exact steps involved in the dynein crossbridge cycle are unclear.

(ii) Chemical Processes in Amoeboid Movement

A number of theories have been put forward to explain amoeboid movement. One of the earliest theories was the tail contraction gel⇌sol theory (Mast, 1926). The sequence involves attachment of a pseudopod to the substratum, gelation of the sol protoplasm at the anterior end, solation of the gel plasma at the posterior end and the contraction of the posterior plasmagel. At the time when this theory was proposed there was a paucity of suitable biochemical techniques to provide experimental evidence to support it at the molecular level. Other theories followed, and one, the frontal contraction hypothesis, which suggested that the motive force for pseudopodial extension was located in the pseudopodial tip, began to gain general acceptance as a quantity of biophysical evidence began to verify it. The endoplasm can be drawn forwards by a tensile force if the cell protoplasm posseses viscoelastic properties. The viscoelastic properties of endoplasm have been demonstrated by polarisation microscopy (Allen, 1972). The endoplasm of amoebae has been shown to be birefringent, and further, the birefringence can be changed by the manipulating of the tension applied to the endoplasm at the tip of a pseudopodium. Birefringence was attributed to strain indicating the viscoelastic properties of the endoplasm. Other experiments conducted by Allen et al. (1971) demonstrated that high negative pressure gradients, in excess of those in the motive force, applied to the tip of one pseudopod of *Chaos chaos*, did not prevent the animal from extending other pseudopodia, thus totally discrediting the tail contraction gel⇌sol theory.

The movement and contractility of amoebae is under direct chemical control. The isolated cytoplasm of *Chaos chaos* has been made to contract by free calcium ions above a concentration of $7.0 \times 10^{-7} M$ Ca^{2+}. The cytoplasmic contractility was cycled through rigor or stabilised contracted and relaxed states repeatedly by simply manipulating the free calcium and ATP concentrations. During the transition from a stabilised state to a relaxed state, a loss of viscoelasticity was demonstrated by strain birefringence (Taylor et al., 1973).

Contractility is an essential process in the movement of sarcodines, but in addition the viscoelastic properties of the cell and changes of sol to gel, and *vice versa,* are an integral element in the process of locomotion. Although at present we are unable to indicate with any certainty how the rheological cycle from gel⇌sol occurs, there is a distinct possibility that the contractile protein actin and the force-generating, energy-transducing enzyme myosin may be involved.

The axonemal microtubules in the axopodia of heliozoans were at one time thought to be involved in mediating movement. Edds (1975a) demonstrated that the axonemes, and other microtubules, played no role in particle motion in the axopodium. He created an artificial axopodium in a heliozoan cell using a fine glass needle of a diameter equivalent to a normal axoneme. Typical particle movement occurred in the cytoplasm of the artificial axopodium. When the cell was treated with the microtubule-inhibitor colchicine, sufficient to cause the collapse of the normal axopodia, the particles in the artificial axopodium continued to move at an almost normal rate. Having proved that the axoneme was not responsible for particle movement, particularly in the cortex and axopodia, Edds (1975b) went on to investigate the cytoplasm for evidence of contractile behaviour and linear elements other than microtubules. Contraction was induced in isolated cytoplasm in calcium ion concentrations above 2.4×10^{-7}M. The addition of 10^{-3}M ATP brought about a rigor to relaxation process. Edds (1975b) identified two types of filaments in the cytoplasm, first, very thin filaments which were identified as actin, and secondly, other thicker filaments which were not dissimilar to the myosin of other amoebae, suggesting possibly the presence of myosin aggregated in an unfamiliar form. It would appear that a contractile system may be responsible for all types of motility performed by helizoans. The contractile system effects locomotion by bending the axopodia against the resistive forces of the microtubular axonemes. The movement of particles in the cytoplasm of the axopodia may be accounted for by localised contractions of the cytoplasm. Clearly there is considerable scope for further investigation into the chemical mechanism of movement in helizoans.

Microtubules are an important element in the reticulopodial network of the foraminiferans, and recent work has shown the presence of other microfilaments in the reticulopodia. Initially it was thought that these microfilaments might be composed of actin, but it now seems unlikely that this is the case (Travis and Allen, 1981). Observations of interactions between cytoplasmic particles and microtubules and 'sliding' and 'zipping' of microtubules (Allen, 1981) indicate that in this group of sarcodines the microtubules do play an active role in motility. Allen (1981) has suggested that the foraminiferans may offer a unique opportunity to study a type of movement which is dependent on the interaction of two forms of linear elements.

D. Forms of Locomotion

The structural proteins, contractile proteins and enzymatic proteins of the Protozoa bring about gross movements of locomotory appendages, whether cilia, flagella or various types of pseudopodia, resulting in the overall movement of the protozoan cell. The co-ordination of locomotory structures is essential in any organism if movement is to be orientated in a specific direction. Efforts to explain this aspect of protozoan motility have been largely directed towards the flagellates and ciliates. There is little information on the co-ordination of movement in sarcodines.

The ciliates, with their complex ciliary organisation, have fascinated protozoologists engaged in motility research. The cilium is a widely occurring structure in the living world, not only providing a means of locomotion for lower organisms such as Rotifera, Gastrotricha and Protozoa, but also appearing commonly in many tissues in Metazoa from advanced phyla. The widespread occurrence of cilia, and the fact that the 9 + 2 structure of the tubules in the ciliary axoneme is universal, has led to a vast literature on the subject. Although much of the literature does not pertain directly to Protozoa, much of it does have a value in understanding the beating patterns of cilia and flagella in protozoan cells. However, unlike most metazoan cilia, the stroke of cilia in protozoans is more versatile, because instead of being restricted to one fixed plane, protozoan cilia can direct their stroke in almost any plane perpendicular to the cell surface. This characteristic is essential to any organism which requires to move in variable directions.

(i) Locomotion in Ciliates and Flagellates

The movement of both cilia and flagella involves the passage of waves along the organelle axis from one end to the other. The waves propagated may be planar or three-dimensional and consequently the sequence of shapes adopted by the cilium or flagellum during the passage of a wave will vary.

Most flagella move in a planar mode only, but in some species movement of the flagella is helical. The amplitude and wavelength of planar waves are quite variable among different flagellate species, and apparently bear no relationship to the length of the flagellum. Although in some cases the amplitude and wavelength of the wave remain constant during passage along the flagellum, more often they both vary, increasing during the propagation of the wave (Sleigh, 1974). In almost all species the waves are propagated distally, so that characteristically the wave is initiated at the flagellum base and progresses along the flagellum in one plane increasing in both amplitude and wavelength during the progression.

In some flagellates movement is complicated by the presence of flagellar appendages, usually mastigonemes, which are a few micrometres long and about 100-200 Å in diameter. In *Ochromonas*, for example, the propulsive thrust is produced by the action of the mastigonemes rather than by the movement of the smooth flagellum. As the flagellum generates waves from its base to its tip, the mastigonemes move passively in such a way as to produce a force acting on the flagellum in the same direction as the wave motion (Figure 4.6). At the crest there is a backwards movement of the mastigonemes relative to the waves. In the trough, however, any influence of the mastigonemes is dissipated by the interference of adjacent mastigonemes (Holwill, 1966).

Three-dimensional or helical waves occur in *Euglena*. The undulations are propagated distally and progress to the flagellar tip. As a result of the helical flagellar beat, the cell is inclined at 30 degrees to its axis of progress, its anterior and posterior ends transversing spiral paths of different radii (Holwill, 1966). The biflagellated dinoflagellates, e.g. *Ceratium*, have flagella performing different waveforms (Figure 4.7). The longitudinal flagellum performs a planar wave, while the transverse flagellum, which lies in a transverse groove, propagates helical waves (Jahn and Bovee, 1964). Other biflagellate species, like *Chlamydomonas*, are believed to effect a breast-stroke movement, helical waves progressing along each flagellum. Usually a species is capable of producing one type of waveform only or, as in the case of dinoflagellates, each flagellum produces one waveform, but some species have been reported as having the capability of generating different types of flagellar motion. In *Peranema trichophorum* proximally-directed helical waves causing the organism to move forwards, flagellum first, have been described (Jahn *et al.*, 1964), as well as distally propagated waves, which may be accompanied by body contortions (Jahn *et al.*, 1963). This latter flagellar movement is rather irregular and may well be part of an avoiding reaction.

Analysing the pattern waves in the cilia of ciliated Protozoa has not

Figure 4.6: The Movement of
Flagellar Mastigonemes.

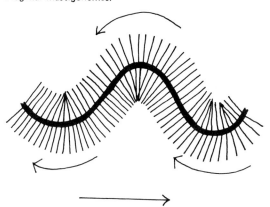

direction of movement

Figure 4.7: Movement in
Ceratium by the Transverse and
Longitudinal Flagella.

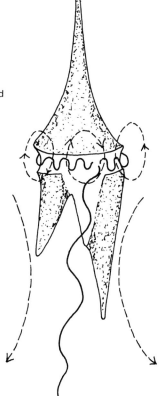

proved an easy task, because cilia are numerous on each cell, each cilium is in close proximity to its neighbours and the form of beat is more complex than that of flagella. The development of rapid fixation techniques has allowed microscopical examination of preserved ciliary waveforms and this showed that the beat of protozoan cilia was not as simple as had been assumed. Instead of occurring in one plane, a three-dimensional pattern emerged. The effective stroke may be planar, but in the recovery stroke the cilium sweeps out to the side, thus creating an overall beat with a three-dimensional pattern (Machemer, 1970).

Since large numbers of cilia occur on each ciliate organism, an effective system co-ordinating the beating of cilia is necessary. The co-ordination of the cilia into metachronal waves was at one time attributed to a neuroid conduction process along a ciliary row. As the impulse passed along it was believed to stimulate each cilium in turn but the cilia themselves did not participate in the transmission of the impulse. Sleigh (1956) showed that if neuroid transmission was the case in the peristomal region of *Stentor*, the transmission of the impulse would be of the order of 600 μm sec^{-1}, which is extremely slow when compared to other conducting tissues. The now widely accepted view is that metachronism in ciliates is the result of hydrodynamic forces acting on the autonomous beating of each cilium (Machemer, 1974). During a beat each cilium transports a surrounding layer of the liquid medium. Obviously the viscosity of the surrounding medium will determine the quantity of liquid carried. Between neighbouring cilia there will be an overlapping in water layers transported, and as a consequence interference will occur between the movements of adjacent cilia. In other words there is a hydrodynamic linkage between cilia, which results in the synchronised beat passing along ciliary rows. External factors such as temperature, light and medium viscosity, together with physiological factors controlling the rate of chemiomechanical transducing, affect the frequency of beat, and hence the velocity of a synchronised wave passing along a row of cilia, as well as the direction of ciliary beat.

Different metachronal patterns can be recognised in Protozoa, and classified according to the angle of the power stroke in relation to the direction of metachronal transmission (Knight-Jones, 1954). Where the wave and the power stroke occur in the same direction, the term symplectic metachrony is applied. Antiplectic metachrony involves waves travelling in a direction opposite to the power stroke of the cilia (Figure 4.8). In both cases the waves and the power strokes occur in the same plane. This form of metachrony is designated as orthoplectic.

The term diaplectic is given to patterns of metachrony where the power stroke and the wave occur at right angles to each other. Two forms of metachrony are found in this category. Dexioplectic metachrony, which is very common in ciliates, is shown in Figure 4.9. Here the power stroke is to the right where the observer is looking along the wave direction. The power stroke is to the left of the observer in laeoplectic metachrony.

Figure 4.8: Orthoplectic Metachrony. a: Symplectic, where the wave and the power stroke occur in the same direction. b: Antiplectic, where the waves travel in a direction opposite to the power stroke of the cilia.

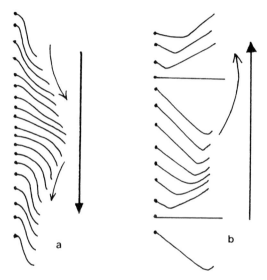

Paramecium shows dexioplectic metachrony and executes a helical swimming path such that the propulsive forces of the cilia must be directed more or less parallel to the wave fronts (Machemer, 1974). *Stentor*, which spends a considerable time attached to a substrate by means of a holdfast, is also capable of free swimming. The body is covered by ordinary cilia, but around the adoral region is a membranellar band of long compound cilia. When swimming, *Stentor* describes a left-hand helical path, during which the membranellar band shows a series of metachronal waves which are propagated in the gullet region and travel around the band. The form of metachronism exhibited by the membranellar band is dexioplectic (Sleigh, 1962). Observations indicate that the body cilia also perform a dexioplectic pattern of metachrony (Machemer, 1974).

Figure 4.9: Diaplectic Metachrony of which Dexioplectic Metachrony Shown Below is Common in Ciliates. Here the power stroke is to the right where the observer is looking along the wave direction.

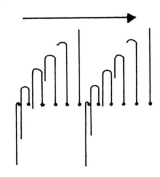

The patterns of metachrony have been described in only a limited number of ciliates, but it is clear that dexioplectic metachronism is the common pattern among these advanced protozoans. As one surveys those Protozoa dependent on flagella and cilia for motility, the pattern changes from the simpler simplectic and antiplectic form into the more complex dexioplectic and laeoplectic form through the progression from Mastigophora to the more advanced Ciliophora. Machemer (1974) suggests that dexioplectic metachrony may have developed from symplectic forms by a progressive polarisation of the counter-clockwise flagellar beating, accompanied by a reduction in the length of the shaft.

(ii) Contractile Movements in Ciliates

In addition to normal ciliary-induced movements, there are some ciliates capable of rapid contractions. Among these are members of the genera *Stentor* and *Spirostomum*. *Stentor* is capable of rapid contractions from the elongated trumpet-shaped form into a pear-shaped form, usually in response to mechanical or chemical stimuli. Studies have shown the presence of two distinct cortical fibre systems which may effect contraction. One of these is microtubular in composition which has been designated as 'km' fibres, while the other consists of fine fibres and has been designated as 'M' fibres (Randall and Jackson, 1958). The behaviour of the M fibres indicates that they are contractile elements. Bannister and Tatchell (1968) suggested that the diffusion connections between the 'M' fibres and the overlying kinetosomes and pellicle could constitute structural attachments which cause changes in the longitudinal dimensions of the cell. In the anterior part of the cell the 'M' fibres form a nearly continuous sheet around *Stentor*. When

Stentor coeruleus contracts, distinct longitudinal bands are apparent consisting of alternate clear and blue pigmented stripes, and the pellicle shows considerable bulging in the striped region. Pellicle bulging may be induced by lateral contractions of the 'M' fibre material, which draws adjacent 'km' fibres together.

The evidence is not entirely conclusive but similar sheets have been described in *Spirostomum* (Finlay *et al.*, 1964). The contractions in *Spirostomum* can be induced experimentally by electrical stimuli when the concentration of calcium in the medium is $10^{-5}M$ or more (Ettiene, 1970). However, the contraction will still occur when the ciliates are transferred to a calcium-free medium, which indicates that the calcium involved is endogenous and released in required amounts to the contractile elements, from depots in the cell. Ultrastructural investigations have revealed membrane-limited vesicles in close association with the subcortical filamentous network and mitochondria. Ettiene (1970) suggests that the mechanism responsible for contraction in *Spirostomum* is analagous to that which operates in striated muscle. Under proper ionic conditions the vesicular network depolarises and releases calcium, which initiates the generation of tension by the contractile machinery of the cell.

(iii) Locomotion in Sarcodina

The frontal contraction theory for explaining amoeboid movement is now well accepted (see Section C(ii)), and also accounts for the motility of naked and many testate amoebae. The testate and shelled species have constraints imposed on locomotion as a result of the weight, and very often the shape, of skeletal structures.

In testate amoebae, of which *Difflugia* is an example, the animal moves by extending cylindrical pseudopodia which attach to the substratum. The pseudopodia then retract forcibly pulling the shelled body forward. There appears to be some element of competition between attached pseudopodia where they are extended at approximately 180 degrees to each other. The least firmly attached pseudopod is detached by the contractile action of the more firmly attached one, and this in turn determines the direction of movement. Closely-extended pseudopodia, on the other hand, very often fuse when they make contact with each other (Wohlman and Allen, 1968).

Foraminiferans move by extending slender filapodia that may reach several millimetres in length in some species, e.g. *Allogromia*. The extending filapodia branch and fuse with each other so that there is a continuous anastomosing reticulopodial network. Within the network bidirectional streaming of particles and cytoplasm is continuous. The reticulopodial pattern contains apparently organised microtubules, which mirror the branching patterns of individual pseudopodia.

The Heliozoida, which are often called the sun organisms because they resemble a stylised representation of the sun, have numerous long stiff protoplasmic extensions, the axopodia, supported by axonemes composed of microtubules. Heliozoans move very slowly, rolling along by shortening and lengthening of the axopodia. The forward axopodia lengthen and become attached; meanwhile the posterior axopodia detach and retract. Some form of co-ordination must exist between axopodia in order to bring about coherent pattern of axopodial lengthening and shortening (Tilney and Porter, 1965). At present no obvious mechanism is apparent.

E. Factors Influencing Speed of Movement

Among the ciliates and flagellates the speed of movement can be fast in relation to size, particularly in the smaller species. Larger ciliate species and the sarcodines progress from one place to another, at a relatively sedate pace. As one might expect, environmental factors, both chemical and physical, impose an effect on locomotion, just as they do on other aspects of protozoan physiology. Unfortunately, there is a paucity of information on the impact of environmental regime on locomotion, which might usefully shed light on motility in a range of natural habitats where protozoan communities are found. The following information must be viewed bearing in mind its limited applicability.

(i) Flagellates

In the flagellates light, temperature and pH have been shown to influence the speed of movement. In *Peranema* rate of movement between 14°C and 28°C increased from around 13.65 μm sec^{-1} to 41.21 μm sec^{-1}, while at 34°C it decreased to around 26.06 μm sec^{-1} and then increased again to 53.7 μm sec^{-1} at 36°C. The magnitude in the rate of increase was greatest at lower temperature intervals, so that between 14°C and 16°C $Q_{10} = 4.23$, whereas between 30°C and 32°C $Q_{10} = 0.83$ (Shortess, 1942). The fluctuation in speed between 28°C and 36°C is suggestive of physiological stress. As the organisms approach their thermal limit, their locomotory behaviour is likely to become erratic. A close examination of the frequency of beat supports this view. Holwill and Silvester (1965) found that flagellar beat increased from three beats per second to 24 between 4°C and 28°C, with a more or less constant wavelength and amplitude, but at temperatures between 30°C and 35°C beating became erratic, and above 35°C localised spasmodic bending of the mid-portion of the flagellum occurred, and waves were not propagated.

Light is an important environmental factor for autotrophic flagellates. As one would expect, these protozoans orientate towards light of moderate intensity, but intense light produces a negative phototaxis. There may be a relationship between light and temperature in controlling the speed of flagellar beat in some flagellates. *Peranema* shows no response in terms of rate of movement when subjected to light of low or medium intensity, but light of high intensity causes a reduction in flagellar beat above 14°C and increases it below 14°C (Shortess, 1942).

pH has been implicated as a factor influencing movement. In *Euglena* a forwards swimming rate of 36.9 μm sec^{-1} at pH 4.8 was found, increasing to an optimum of 59.7 μm sec^{-1} at pH 7.0, thereafter decreasing at higher pH. *Chilomonas* has a faster rate of movement. At pH 4.8 the speed is 90.3 μm sec^{-1} increasing to 162.8 μm sec^{-1} at pH 7.0, declining to 74.4 μm sec^{-1} at pH 8.5 (Lee, 1954).

(ii) Ciliates

Changes in the temperature of the aquatic medium affect the velocity of wave transmission of cilia, without changing the orientation of the metachronal pattern. Low temperature reduces beat frequency. In *Stentor*, for example, a decrease in temperature from 36.0°C to 5.3°C decreased beat frequency from 32.8 beats sec^{-1} to 10.25 beats sec^{-1}. High temperatures, on the other hand, produced a decrease from both the maxima of beat and wave velocity, with the cilia showing a tendency to beat intermittently (Sleigh, 1956). The maxima occur at temperatures higher than those normally encountered in the natural environment. As a general rule the rates of locomotion of most free-swimming Protozoa begin to decrease at about 30°C, usually becoming abnormal and ceasing between 30°C and 40°C (Jahn and Bovee, 1967). There are of course exceptions, since some Protozoa have become physiologically adapted to abnormal high-temperature environments.

Ciliate locomotion is also influenced by the concentration of hydrogen ions in the medium. There are numerous reports, but many are of limited value because the past culture history of the organisms is unknown. Various species of *Paramecium* have been reported as showing maximum swimming speed in the slightly alkaline environment of pH 7.5 (Lee, 1956). Dryl (1959) found that *Paramecium caudatum* produced its fastest swimming speed at pH 5.3 when previously acclimatised to pH 7.1 for 24 hours.

(iii) Sarcodina

Among free-living Protozoa the sarcodines are the slowest moving, particularly those species endowed with skeletal structures. The slow erratic dance movement of helizoans achieves speeds varying from 5 to 100 μm sec^{-1} (Watters, 1968), while *Difflugia* has speeds of cytoplasmic streaming of 2-15 μm sec^{-1} (Wohlman and Allen, 1968). The large naked species *Amoeba proteus* has had various speeds reported for its locomotion, ranging from 4.69 μm sec^{-1} (Mast and Prosser, 1932) up to 24.17 μm sec^{-1} (Jahn and Bovee, 1967). The lower speed occurred at 30°C, whereas the faster speed was achieved at 22.5°C. The rate of movement at the higher temperature may possibly be verging on the abnormal, since this temperature is outside the usual environmental range of *A. proteus*.

Unlike the majority of ciliates and flagellates which are able to swim freely suspended in their aqueous environment, many sarcodines adopt locomotory processes requiring contact with a substrate. Because of this, the type of substrate may impose constraints on the rate of movement which can be achieved, particularly in soil-dwelling amoebae. King *et al.* (1981) looked at substrate specificity in *Naegleria gruberi*, a common soil species. They measured the rate of locomotion on glass, plastic, agar and oil in differing NaCl solutions in the range 0-10 mM, and found no significant impact on the type of substrate on movement. The chemical make-up of the medium, however, did impose a constraint on movement. In pure water speed ranged from 0.16 to 0.22 μm sec^{-1} increasing to 1.11-1.26 μm sec^{-1} in 10 mM NaCl. The lack of substrate specificity is not surprising given the chemical heterogeneity of the soil habitat.

5　TROPHIC RELATIONS OF PROTOZOA

A. Introduction

Free-living Protozoa occupy a range of trophic levels. Those which feed on algae, both the unicellular and filamentous varieties, are primary consumers in the herbivore food chain dependent on fresh autotrophic organic matter. Other protozoans exploit bacteria as a food source and as microbivores perform a role in the food chain based on dead organic matter. Both these trophic groups of Protozoa, and other small elements in the micro- and meiofauna, are exploited by predacious protozoan species. The Protozoa form an element in the larger micro- and meiofaunal communities of soils and aquatic environments, but on a smaller scale the various protozoan groups themselves constitute a community with a defined trophic hierarchy (Figure 5.1).

The decomposition food web based on dead organic matter is extremely important in nature, because it is responsible for recycling carbon, phosphorus, nitrogen, sulphur and other essential elements in the living world. The pool of dead organic matter which the decomposers exploit, is derived from primary production and is composed of dead plant tissue, the faeces of heterotrophs, their exudates and exuviae and body remains where death is non-predatory. Thus two basic food chains can be distinguished in nature, one based on fresh plant tissue and the other on dead organic matter, the former providing an energy source for the latter; the decomposers recycling the essential nutrients required for photosynthesis.

The majority of Protozoa are microbivores exploiting the decomposer bacteria as a source of energy, transforming some of the assimilated bacterial production into protozoan tissue and some into energy for metabolic processes and movement. The Protozoa in turn become potential energy for a variety of predators including carnivorous Protozoa, copepods and other meiofauna. The environments inhabited by protozoans are often heavily dependent on the decomposer food chain. Many freshwater and estuarine aquatic habitats, particularly the benthic zones, are part of so-called subsidised ecosystems. These are ecosystems,

Figure 5.1: The Trophic Structure of a Protozoan Community.

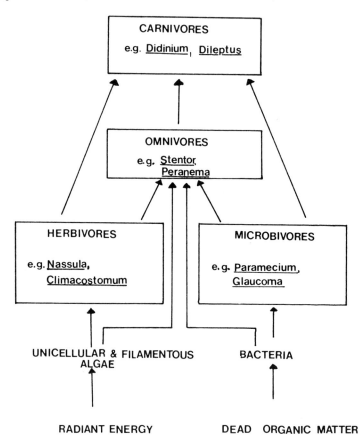

which although performing endogenous primary production, much of which eventually finds its way into the pool of dead organic matter (i.e. autochthonous dead organic matter), also receive an input or subsidy of dead plant tissue from surrounding terrestrial or semi-terrestrial ecosystems (i.e. allochthonous dead organic matter). Soils and litter layers are also important protozoan habitats and represent the area of terrestrial ecosystems where the decomposition processes are performed.

The obvious ecological role of microbivore Protozoa is the transfer of bacterial production through protozoan cellular material as potential energy to successive trophic levels. However, it is now becoming increasingly clear from studies on nutrient recycling and bacterial growth dynamics, particularly in the soil, that the contribution made by Protozoa may be greater and more complex than we had previously assumed. There is now evidence that Protozoa may stimulate bacterial growth and hence the rate of decomposition by various means. In addition it seems that Protozoa play a significant role in the recycling of essential nutrients, especially carbon and phosphorous.

The efficiency and rate at which protozoans transfer energy from one trophic level to another, is determined from physiological studies on species populations. The physiological data on growth, respiration and feeding are transformed into units of energy, so that the energy intake and partitioning of energy within the organism can be considered. This energetics approach integrates the various aspects of an organism's physiology, so that the overall functioning of an organism as an element in a population or a community can be ascertained. The flow of energy through a species or population can be expressed simply in the energy budget equation, which was proposed by Ivlev (1938):

$$C = P + R + F + U$$

where C = the energy ingested as food; P = energy produced as new tissue and reproductive products; R = heat dissipated during respiration or respiratory energy loss; F = egested energy; U = nitrogenous excretion.

$$A = P + R$$

where A = energy assimilated or utilised.

In protozoans P is solely growth in the normal asexual life-cycle, because the whole body is the reproductive product, passed on usually as two new individuals to the succeeding generation. Egestion and excretion are for obvious reasons logistically difficult to quantify in such small organisms, so these are usually calculated by subtraction from the other empirically-determined parameters of the energy budget.

The efficiency by which an organism uses the energy it ingests for production and metabolic function is expressed by a series of ratios which are termed coefficients of efficiency:

$$\text{assimilation efficiency} = \frac{A}{C} \text{ or } \frac{P + R}{C} \times 100$$

This indicates what proportion of the ingested food or energy is actually used by the organism for production and respiration.

$$\text{gross production efficiency} = \frac{P}{C} \times 100$$

which gives us a crude index of how much of the consumed energy appears as production.

$$\text{net production efficiency} = \frac{P}{A} \times 100$$

This indicates the proportion of assimilated energy which is directed into growth and reproduction. If the proportion of the assimilated energy going into production is known, then it follows that the proportion which is dissipated as heat during respiration can be derived by subtraction.

These various coefficients of efficiency are very valuable when making comparisons of organisms from the same trophic level, and for comparisons between trophic levels. At a community level the components of the energy budget expressed in joules or calories, and the coefficients of efficiency, give us an insight into the overall energy flow and energetic interactions between organisms.

B. Protozoan Energetics

Studies on protozoan energetics are relatively recent, although the physiological ecological approach to the functioning of organisms developed in the 1940s, and gained momentum, so that by the 1960s and 1970s a wealth of information was available on the energetics of invertebrates and vertebrates in the ecological literature. The first energy budget produced for a protozoan was for the amoebae *Acanthamoeba* spp. (Heal, 1967). This energy budget, valuable though it was in leading the way for subsequent protozoan energetics studies, had its limitations, since it applied to one temperature only and suffered from a lack of calorimetric data. Later investigations attempted to integrate the variable of temperature, so that energy budgets could be applied to field populations. The aim of energetics studies is to elucidate the physiological performance of an organism in relation to its food source and its predators in the natural environment. In this respect, protozoan ecology lags behind that of many other invertebrate groups, particularly macrofaunal species. The area is beset by problems, not least of which is our knowledge of the nutritional requirements of most protozoan species. They can be grown in the laboratory in axenic and monoxenic culture, but little is known about protozoan feeding preferences in the wild, or about how the elements of competition for food on an intra- and interspecific basis affect energy consumption and other aspects of the physiology. As Lee, J.L. (1980) points out, our understanding of energy transformations in food webs involving Protozoa require an elucidation of two components related to food quality: first, informa- tion on the molecular constitution of the prey, and secondly, the ability of Protozoa to recognise and utilise it. Thus while we are able to demonstrate how food quality and quantity affect growth and reproduction in many species, we can only place this information into the context of the natural community to a limited extent.

The reasons for the paucity of information on protozoan energetics are several. Until recently the smaller animals in the ecosystem were mentioned only in passing, if at all, and dismissed as probably contri- buting little to the overall functioning and energy flow of aquatic and terrestrial communities, and hence there was no need to study them in detail. Protozoan physiology and calorimetry are fraught with technical difficulties imposed by the small size of these organisms. The ideal invertebrate for an energetics investigation is a macrofaunal mono- phagous herbivore, with a one-year life-cycle and recruitment at one specific period only. The Protozoa are neither macroscopic, nor mono- phagous on one species of bacterium or alga, and they have continuous recruitment over a short period. Furthermore, they frequently live in communities with high species diversity.

(i) The Energy Equivalents of Protozoa and their Food

In order to compose a picture of energy flow through the physiological function of a population or community, it is necessary to acquire data on the energy value of organisms and their food. The energy content of such materials is usually determined by means of microbomb calori- metry (Phillipson, 1964).

Table 5.1 shows the energy values of a range of protozoan species derived from a number of studies. The energy values derived for various ciliate species do not differ radically, and it is not unreasonable to apply such values as conversion factors to other ciliate species. The low value for *Noctiluca* is attributable to a relatively high volume of cell sap and probably a water-bound component included in some of the

Table 5.1: The Joule-equivalent Values of Protozoa Derived from Bomb Calorimetry

Species	$J\ mg^{-1}$ Ash-free Dry Wt	Author
Ciliata		
Tetrahymena pyriformis	24.85 ± 0.84	Slobodkin & Richman (1961)
Tetrahymena pyriformis	21.3 ± 0.3	Finlay & Uhlig (1981)
Tetrahymena pyriformis	19.8 ± 0.2	Rogerson (1979)
Paramecium caudatum	23.5 ± 0.7	Finlay & Uhlig (1981)
Euplotes spp.	19.3 ± 0.7	"
Colpidium campylum	20.15 ± 1.5	Laybourn & Stewart (1975)
Sarcodina		
Amoeba proteus	17.51 ± 0.3	Rogerson (1979)
Mastigophora		
Ceratium hirudinella	23.2 ± 0.8	Finlay & Uhlig (1981)
Noctiluca miliaris	6.6 ± 0.6	"

samples (Finlay and Uhlig, 1981). There is obviously a need for more detailed information on flagellates and on amoebae where data are limited. The energy content of protozoan cells is within the range reported for a wide spectrum of invertebrate organisms showing energy values within the range 17.58-28.46 $J\ mg^{-1}$ ash-free dry weight (Prus, 1970).

The bacterial food sources of many protozoans show variable energy values. Laybourn and Stewart (1975) report *Moraxella* isolated from fresh water as having an energy content of 20.36 ± 0.4 $J\ mg^{-1}$. Other freshwater species isolated from *Paramecium* were found to have an energy content of 21.0 ± 0.4 $J\ mg^{-1}$, while marine bacteria have a calorific value of 5.5 ± 0.8 $J\ mg^{-1}$ (Finlay and Uhlig, 1981). The lack of information on the food of protozoans emphasises the problems outlined earlier.

(ii) Energetics

Complete energy budgets related to variables such as food concentration and temperature are few, although there are more examples of what may be described as partial energy budgets covering feeding and production. Those complete budgets which are available do provide some insight into the overall efficiency of exploitation and use of resources by protozoans. Table 5.2 shows some energy budgets for two very different protozoans feeding on *Tetrahymena* in an experimental situation. One of the species, *Amoeba proteus*, moves along a substrate engulfing the food it encounters, while the other species, *Stentor coeruleus,* is a large, mainly sedentary ciliate which actively draws water over the oral region where it filters out food items. In each case a single *Amoeba* or *Stentor* was placed in a volume of 500 μl of medium with differing concentrations of prey. In the case of *Amoeba* the density of the prey was considerably greater than that available to *Stentor.* The higher prey concentrations offered to *Amoeba* may not reflect the natural situation, but do result in higher levels of consumption. Indeed the level of consumption continued increasing up to prey densities of 4,000 per *Amoeba* (Rogerson, 1981). Given the much lower densities of food available to *Stentor,* comparatively high levels of consumption were achieved, especially when one considers that the energy budget for *Stentor* spans a shorter period because the generation times are shorter than those in *Amoeba* (see Table 5.2). In other words the rate of consumption is apparently higher in *Stentor* on lower food densities than in *Amoeba proteus.* This disparity is a function of the mode of feeding in each species. *Stentor* filters the medium in which it lives for food. The rates of filtering which can be achieved by ciliates are high. It has been estimated that a ciliate can clear a volume of water equal to 20,000 times its own volume in one hour. Put in perspective, this is 20 times higher than lamellibranch molluscs (Fenchel and Jørgensen, 1977). *Amoeba proteus* has a less efficient means of acquiring energy and depends upon random contact with its prey as it moves along, and thus much lower rates of consumption are achieved in relatively high prey densities.

Having consumed energy, an organism then extracts and utilises the energy available to its digestive capabilities from the food, and partitions the assimilated energy into production and maintenance. The efficiency by which energy is derived from ingested food is given by the assimilation efficiency (A/C). The figures in Table 5.2 show that *Stentor* possesses higher assimilation efficiency, taking more of the available energy from the same type of food than *Amoeba.* The reasons are difficult to pin-point, but the differences may be a function of a more efficient system of digestion and absorption in *Stentor.* This species has a higher overall production and higher rates of production than *Amoeba,* which is apparent from the quantity of μJ produced and the difference in the generation times. Put another way, *Stentor* grows to a larger size in a shorter time and produces more generations in a

Table 5.2: Generation Energy Budgets for Protozoans. Food source *Tetrahymena pyriformis*. Parameters of the energy budget expressed as $\mu J \times 10^3$. See text for explanation of abbreviations

Species	Gen. Time hrs	Temp °C	Food Ratio	C	P	R	F+U	A	$\frac{P}{A}$ %	$\frac{P}{C}$ %	$\frac{A}{C}$ %
Amoeba proteus	111	15	125:1	4.68	1.78	0.59	2.30	2.37	75	38	50
"	68	15	2,000:1	11.21	2.53	0.61	8.06	3.14	80	22	28
"	83	20	125:1	4.46	1.23	0.66	2.56	1.89	65	27	42
"	48	20	2,000:1	5.88	1.75	0.49	3.63	2.25	78	30	38
Stentor coeruleus	43	15	30:1	2.93	2.21	0.06	0.66	2.26	97	75	77
"	40	15	50:1	4.26	2.71	0.06	1.48	2.77	97	63	65
"	42	20	30:1	3.04	2.17	0.09	0.77	2.66	96	71	74
"	39	20	50:1	4.39	3.23	0.12	1.03	3.35	95	74	76

Source: Data from Rogerson (1980, 1981) and Laybourn (1976b).

given time than *Amoeba*. The efficiency of net production (P/A) is higher in *Stentor*, but both species are efficient producers. In each case a large proportion of the assimilated energy is directed into production, and respiratory costs are low. The very low levels of respiratory energy loss in *Stentor* are probably attributable to its largely sedentary habit. These carnivorous protozoans possess net production efficiencies which are considerably higher than those found in metazoan carnivores. A review of production efficiencies in ectothermic metazoans by Humphreys (1979) indicates that non-social carnivorous insects achieve net production efficiencies around 56 per cent, while other metazoan invertebrate carnivores are lower, averaging 28 per cent.

Net production efficiencies among the small number of bacterivorous Protozoa which have been investigated show considerable variation. *Strombidium* sp., *Tiarina fusus* and *Diophrys* sp. achieve levels of efficiency of 2-6 per cent, 15-42 per cent and 14-40 per cent respectively at temperatures ranging between 22°C and 25°C (Klekowski and Tumantseva, 1981), while another ciliate, *Spirostomum ambiguum*, has a net production efficiency of 55 per cent (Klekowski and Fischer, 1975). The bacterial feeding flagellates *Ochromonas* sp. and *Pleuromonas jaculans* are capable of net production efficiencies around 60 per cent at 20°C (Fenchel, 1982). *Strombidium* would appear to be exceedingly inefficient, but this may be a function of the experimental temperature; it may be a species better adapted to lower temperatures. The other species are capable of efficiencies within and above those reported as characteristic of metazoans (Humphreys, 1979). However, it should be noted that Humphreys (1979) did not include bacterivores among the trophic groups in his review, probably because the literature on such feeders is sparse. The efficiency at which bacterivorous protozoans convert assimilated energy into production is comparable to that displayed by the bacterial feeding nematode *Plectus palustris*, which is capable of net production efficiencies in the range 51.2-73.7 per cent at 20°C, varying in relation to developmental stage and food concentration (Schiemer *et al.*, 1980).

It is interesting to speculate on how the relationship between respiration, production and assimilation, and consequently the efficiency of energy utilisation by an organism, has changed during the course of evolution. Phillipson (1981) suggests that within multicellular ectotherms production efficiency fell progressively during evolutionary advancement, while maintenance costs, as a proportion of total assimilation, increased. The development of homeothermy was accompanied by a further fall in production efficiency and an associated increase in maintenance costs. At present there are insufficient data on unicellular organisms to draw any firm conclusions, but the high net production efficiencies displayed by many of the protozoans so far studied would suggest that Protozoa may be among the most efficient heterotrophic organisms, and that the evolution of a multicellular level of organisation and associated morphological and physiological complexity, has resulted in a decrease in the efficiency of production.

Production studies related to food consumption give us a gross estimate of conversion efficiency (P/C) of ingested energy into production or yield. The gross production efficiency is a useful ecological index of the quantity of energy transferred as yield by an organism from the trophic level it exploits as food to its own predators in the trophic level above it. Those bacterial and fungal feeding Protozoa which have been studied in this respect have varying gross production efficiencies. Heal (1967) found *Acanthamoeba* sp. fed on yeast cells to attain a gross conversion efficiency of 58 per cent at 25°C. *Tetrahymena pyriformis* grown on bacteria achieved a gross production efficiency of 50 per cent at the same temperature (Curds and Cockburn, 1968). Higher levels were achieved by *Colpoda steini* where gross production was 78 per cent of consumption (Proper and Garver, 1966), while lower values of P/C ranging from 3 to 11 per cent over 10-20°C have been reported in *Colpidium campylum* feeding on bacteria (Laybourn and Stewart, 1975). The bacterivorous flagellates *Ochromonas* sp. and *Pleuromonas jaculans* achieved yields of 34 per cent and 43 per cent at 20°C respectively (Fenchel, 1982). The wide variations on these gross production efficiencies for Protozoa are probably partly a function of the temperature at which experiments were conducted – generally the efficiency of production declines with decreasing temperature – but other factors may be involved. Production may also vary in relation to food concentration (see Chapter 3, Figure 3.10), because the quantity of available energy may limit the rate of ingestion, but in addition the energy value of the various bacterial and fungal food sources may vary. We have only limited information available to us on the energy content of these materials (see Section B(i) above).

Of necessity, the foregoing energetics studies have been performed in the laboratory, but production studies for field populations have been attempted, usually combining aspects of laboratory-determined energetics with field data in a number of ways. One of the earliest studies which adopted this approach was carried out by Fenchel (1975) on an arctic tundra pond. Fenchel estimated carbon flow through the most important groups of Protozoa in the pond, giving a minimum estimate of standing stocks as well as transfer rates. The rates of protozoan digestion of bacteria were determined in the laboratory at a series of temperatures. The digestion rates were multiplied by the number of food vacuoles in freshly collected Protozoa to give an estimate of grazing rates in the field. Zooflagellates were quantitatively the most important protozoan group. During 24 hours in the summer months the zooflagellates consumed 15 mg C m^{-2}. The calculations showed that the zooflagellates consumed a number of bacteria equal to their body weight in 24 hours, whereas ciliates grazing on bacteria ingested the equivalent of 25 per cent their own biomass in the same period. These rates of bacterial grazing are high and serve a valuable role in the saprovore food web (see Section below).

In the benthos of many temperate lakes ciliates tend to be the dominant protozoan group. Finlay (1978) estimated production and respiration for the ciliate community of a small eutrophic unstratified lake in Scotland, by combining data on the reproductive and respiration rates of a range of ciliate species (see Figures 3.8 and 3.15) with seasonal field data on numerical density and species composition. The seasonal patterns of production and respiration decreased with increasing depth, Figures 5.2 and 5.3. Mean production ranged from 40 to 345 J cm^{-2} year^{-1}, while respiration ranged from 4 to 16 cm^{-2} year^{-1}. These high levels of production are at their maximum in the summer months. In stratified eutrophic lakes the picture is quite different, for here production is depressed in the summer by the conditions of low temperature and anoxia which prevail in the benthic zone (Figure 5.4). The differences in production between benthic protozoan communities in stratified and unstratified lakes is obvious from a comparison of Figures 5.2 and 5.4. Studies have shown that many of the ciliates in the benthic zone of stratified eutrophic lakes migrate vertically in the water column out of the anoxic conditions of the sediment surface and lower waters (Bark, 1981; Finlay, 1981).

Protozoan communities from running water have been subject to energetics studies, notably by Schönborn (1977, 1981a, b). He exposed testate amoebae on slides in the River Saale in Germany, and derived measurements of production, density and biomass over a period of weeks. Consumption was estimated by laboratory determinations. Production by Testacea in the Saale was of the order of 2.6 kJ m^{-2} per annum, giving a production-to-biomass ratio (P/B) of 16.8. The levels of consumption were estimated as 2,444 mg diatoms m^{-2} year^{-1} (Schönborn, 1981a). In another study on ciliated Protozoa in a small brook, annual production was 4,219 mg m^{-2} with an extremely high

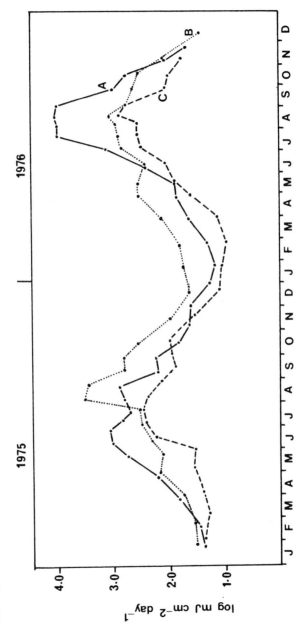

Figure 5.2: Production by the Ciliate Community of a Small Eutrophic Unstratified Lake at Three Sites (A, B & C) of Increasing Depth, A Being the Shallowest Site.

Source: Finlay (1978), with the permission of Blackwell Scientific Publications Ltd.

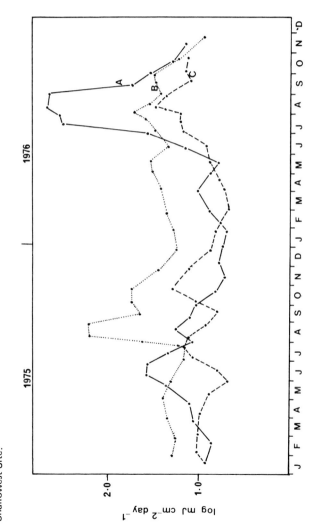

Figure 5.3: Respiration by the Ciliate Community of a Small Eutrophic Unstratified Lake at Three Sites (A, B & C), A Being the Shallowest Site.

Source: Finlay (1978), with the permission of Blackwell Scientific Publications Ltd.

Figure 5.4: Production by the Ciliates of a Small Eutrophic Stratified Lake, Esthwaite in the English Lake District. The data shown are for three depths: □ − 15 metres, ○ − 12 metres, ● − 10 metres.

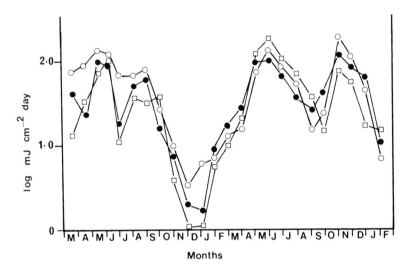

production-to-biomass ratio of 195 (Schönborn, 1981b). The ciliates achieve high levels of consumption, ingesting 14.8 g m^{-2} of bacteria, diatoms and other Protozoa, the largest portion being bacteria.

A river environment is not an optimal one for many Protozoa and one may reasonably expect standing water communities, particularly those tending towards eutrophy, to be more productive. This is borne out by a comparison of the available figures for annual production in lakes and rivers. The production achieved by ciliates in Schönborn's (1981b) study of a river are two or three orders of magnitude lower than those reported for lake ciliates in an unstratified eutrophic lake by Finlay (1978) and in Esthwaite, a dimictic eutrophic lake. The physical and chemical conditions of standing and running waters differ, and this is reflected in the energetics of communities in each type of aquatic ecosystem.

(iii) How Protozoa Maximise Net Energy Returns from their Feeding Behaviour

The acquisition of food or energy requires some expenditure of energy in terms of searching and handling food. For the most efficient functioning an organism must maximise the net energy return from its feeding activity. Optimal foraging theory has been applied to Metazoa, particularly carnivores, but the concept can be applied in a modified form to other types of feeders.

Protozoa fall into different trophic categories (See Chapter 2). Many ciliates and flagellates are essentially filter feeders, drawing the liquid medium over a specific area of the cell and extracting food from the water current. Many other Protozoa, particularly amoebae, are deposit feeders, engulfing particles of food with which they make contact. Carnivorous Protozoa vary in their feeding behaviour; some, such as *Didinium*, actively hunt, but capture is random. The ciliate does not actually pursue a prey, it depends on random contact. Other carnivores, for example the suctorians and representatives of other groups, are ambush predators, waiting for their prey to come to them. Much of the current optimal foraging theory may not be applicable to Protozoa, but since there is evidence of food selection by Protozoa (see Chapter 2), there is scope for considering how some of these single-celled organisms maximise energy return from their feeding behaviour.

Lam and Frost (1976) and Lehman (1976) have applied the principles of optimal foraging to filter feeders. In carnivores it is necessary to consider energy yield per unit handling time of different prey items with differing energy content in different densities, in relation to searching time. In filter-feeding organisms it is the particle density in the medium, the size of the particles in relation to particle selection, and the rate of food processing which are important. Here the filtering rate of the medium by the organism corresponds to the searching time in carnivores. In filter-feeding metazoans, such as bivalves or filter-feeding copepods, the net rate of energy gain equals the rate of energy assimilation, minus the cost of filtering, minus the cost of rejecting unsuitable particles. Townsend and Hughes (1981) have summarised the principles of maximising energy return in filter-feeding invertebrates:

$$Q = E_a - E_f - E_r$$

where Q (= E/T) = net rate of energy gain; E_a = rate of energy assimilation; E_f = cost of filtering; E_r = cost of recycling unwanted particles.

Each component of the equation is complex. The rate of energy assimilation (E_a) is a function of the proportion of particles of any given type which are acceptable for ingestion, the density of any given particle type in the water, the energy value of the particle and the filtering rate. The cost of filtering (E_f) has to take into account the energy expenditure as a result of drag on the filtering apparatus. Rejection energy costs (E_r) are a function of the proportion of any particle type rejected and their density in the medium in relation to the filtering rate. Lehman (1976) plotted filtering rates as a function of particle density. A maximum filtering rate is achieved at a particular particle density of a given particle type; the maximum corresponds to the density of the particle when the gut first becomes full, and thereafter lower filtering rates maintain the gut in a full condition. In nature animals are not usually confronted by one type of particle only, but an array of particles, some of which are suitable for ingestion and some which are not. Those which fall into the category of being suitable as food may have a different energy content. The energy value of any given type of particle may be a function of its size or its digestibility. In order to maximise the energy return from feeding, those particles with the highest energy yield should always be accepted, when their density in the array of particles filtered is proportionally high. However, as the density of energy-rich particles declines in the overall volume of particles being filtered, rejection costs become high relative to the energy gained. When rejection costs are high in filter-feeding metazoans, even non-digestive items should be ingested and passed through the gut along with energy-yielding material (Townsend and Hughes, 1981). The organism must of course be capable of recognising and grading the particles it encounters in terms of their energy value.

Theoretically the same principles are applicable to filter-feeding protozoans. Although there is information available on how feeding currents are produced in ciliates and how much food is consumed per unit time, our knowledge of filtering rates in relation to particle size and concentration is limited. In a number of respects Protozoa appear to be much simpler than many metaozoan filter feeders. Many bacterivorous ciliates appear to discriminate what they ingest on the basis of particle size and shape only, and this is related to their mouth morphology (Fenchel, 1980d). In other words they apparently ingest all particles in the size range within the retention capacity of the oral cilliary organelles, irrespective of the energy value of the particles. There is evidence, however, that some ciliates, such as the large omnivorous species *Stentor,* do discriminate the type of particle ingested (see Chapter 2, Section C). In ciliate bacterivores Fenchel (1980b, d) showed that each species has its own size range of particle, there being a correlation between the minimum particle size retained and the free space between adjoining cilia of the ciliary filters. Rejection occurs in the sense that particles outside the size repertoire of a given species are not retained, so that some energy cost is implied.

It appears that the filtering rate in ciliates is constant irrespective of particle density in the medium (Fenchel, 1980b). The uptake of particles is a linear function of particle density increasing, experimentally, as a linear function of time. The maximum uptake rate is determined by the rate at which particles can be phagocytosed. As particle density in the medium is increased, the filtering apparatus becomes clogged by particles, because the rate at which ingestion can be achieved is limited; thus some of the retained material will be lost from the ciliary apparatus (Fenchel 1980a, b). The energetic cost of overcoming the resistance or drag to the ciliary filter, which presumably is constant since the filtering rate is constant, has been estimated as being very small, of the order of less than 1 per cent of the energy budget of a small ciliate like *Cyclidium* (Fenchel, 1980b). Added to this of course is the energy cost of bringing about ciliary movement.

Do these comparatively simple organisms attempt to maximise energy return from their feeding activity as do many metazoans? The answer is apparently not in most of the ciliate species for which detailed information on feeding activity is available. Any organism which filters at a constant rate irrespective of the density of particles in the medium is energetically inefficient. To be efficient an organism must modulate the filtering rate so that in any given particle density the quantity of material passing over the filtering apparatus to the mouth is the maximum which can be adequately coped with. A protozoan with an apparently constant filtering rate will, in high particle densities, receive too much material to phagocytose and blocking of the filter occurs. Similarly it could be argued that any organism which indiscriminately ingests all particles in a particular size range, is not maximising energy return. A system of recognising and accepting only that material in the array available which gives a good net energy return, is clearly energetically efficient.

Protozoa are very efficient energetically, having high growth rates and low respiratory costs in general. Perhaps there is no real advantage to a unicellular, efficient animal living in a relatively rich food environment in evolving or pursuing a 'strategy' designed to maximise energy return. Such 'strategies' may have evolved in the ecological interactions of higher organisms, in response to longer life-cycles and greater physiological complexity imposing pressure to 'conserve' energy. The size of particles retained by ciliates appears to be fairly small in range. For example, bacterivore species most efficiently retain particles of 0.3-1 μm (Fenchel, 1980a), so that there is a high probability that most of the material ingested will be bacteria, although of course not all the bacteria will be digestible or provide a good net energy return. In other respects ciliates do reduce the energy costs involved in feeding, in that several species have been shown to flocculate bacteria (see Chapter 2, Section B(i)), thus concentrating the food source and rendering it more accessible to grazing.

Ciliates are among the most efficient Protozoa. Less efficient groups may need to modify their feeding behaviour to maximise energy returns. As yet our knowledge of the intricacies of protozoan feeding behaviour under conditions which resemble those of the natural habitat is very limited. We have a wealth of information on the ultrastructure of the structures involved in feeding, and much data on how energy is ingested in relation to environmental physical and chemical conditions, but this is insufficient to consider optimal foraging theory as it may apply to the various protozoan trophic groups. It is an area of protozoan ecological energetics which should be explored, because it sheds more light on an organism's functional interactions with its food sources and on intrinsic energetics.

C. r-K Selection in Protozoan Populations

During evolution, adaptations conferring competitive ability are believed to have evolved with a resulting loss in the innate or intrinsic rate of increase (r_m). This hypothesis, referred to as the r-K selection theory, has been widely applied in animal ecology since it was first proposed by MacArthur and Wilson (1967). Any population in any given environment will have its own characteristic survival rate, growth rate and reproductive rate. Each of these parameters will be partly innate and partly a result of prevalent environmental factors. The ability of an organism to increase, that is its innate or intrinsic capacity for increase (r_m), can be defined under a given set of conditions. Ecologically the hypothesis relates the inner characteristics of the organism to the physical, chemical and biological features of the environment. It has been suggested that some groups of organisms can be recognised as being r- or K-selected (Pianka, 1970).

r-selected organisms are subject to density-independent population control, which results in the population continuously increasing, but rarely reaching its limit. Among Metazoa, such organisms usually have high values for r_m, rapid development, reduced body size, normally a single phase of reproduction, a short life-cycle and reduced competitive ability. Organisms which are K-selected are subject to density-dependent regulation, resulting in the population usually reaching its saturation density. Typically, K-selection involves slow development and a delay in reproduction, which is repeated, together with a large body size and a long life-cycle, usually exceeding one year. Intra- and interspecific competition is marked in K-selected animals. The hypothesis rests on the principle that selection acts reciprocally on the intrinsic rate of increase (r_m) and the population saturation density (K). Thus r-selection under continuous growth should increase r_m, but reduce saturation density (K), whereas K-selection reduces the value for r_m but increases competitive ability and K.

The r-K selection hypothesis has recently been tested on protozoan populations. Luckinbill (1979) set out to examine the effect on the saturation density (K), of selection for higher growth rate, maintaining that the concepts of r- and K-selection apply equally well to microorganisms and metazoans. He tested the hypothesis on four strains of *Paramecium primaurelia,* which represent groups of well-studied Protozoa. Here r-selection was applied to a population by restricting growth densities well below the limit set by the food supply, which if done frequently controls population size independent of density, thus permitting continuous growth at high densities without competition. The prediction that r-selection should reduce the saturation density was not proved. The results showed that r-selection favoured high values of both r_m and saturation density (K). Among the strains of *P. primaurelia* tested, better competitors had reduced saturation densities. Gill (1972), however, found that neither r_m nor K was related to competitive ability among strains of the *Paramecium aurelia* complex. When Luckinbill (1979) considered a range of ciliates of varying cell volumes and biomass, one aspect of the r-K selection hypothesis was apparent in Protozoa: the better competitors were larger species, with high individual population biomass. Large ciliates usually have lower growth rates than smaller species. The poorer competitors were small ciliate species, with lower overall biomass and high growth rates.

A different approach was adopted by Taylor (1978b), who tested the hypothesis with 11 species of bacterivorous ciliates in a small pond, comparing the measured or predicted innate rate of increase (r_m) of a species with its commonness in the field. The prediction was that those species with high values of r_m would be less common, as indicated by the number of samples in which they were found. Taylor used the index $r/\hat{r}$, where r is the experimentally estimated r_m and $\hat{r}$ the r_m predicted by the equation of Fenchel (1968) (see Chapter 3 and

Figure 3.8). A fast-growing species has a high $r/\hat{r}$ value, while slow-growing species have a low value for $r/\hat{r}$. As predicted there was a negative relationship between $r/\hat{r}$ and commonness, the species with a high $r/\hat{r}$ value being less common than those with a low $r/\hat{r}$ value (Figure 5.5). Paradoxically, the species designated as r-selected by $r/\hat{r}$ are mainly large species, but Taylor (1978b) argues that $r/\hat{r}$ is to some extent independent of size, and although size does enter into the calculation, an animal of any size or r_m can be assigned to the K-selected or r-selected category. Physiologically larger ciliates may be better adapted to withstand periods without food, and small species capable of more efficiently exploiting the small-scale spatial heterogeneity that occurs during periods of resource limitation.

Figure 5.5: The Relationship between $r/\hat{r}$ and Commonness among Ciliates.
Cg — *Cyclidium glaucoma*, Cm — *Cinetochilum margaritaceum*, Cco — *Colpidium colpoda*, Cc — *Colpidium campylum*, Gf — *Glaucoma frontata*, Gs — *Glaucoma scintillans*, Hg — *Halteria grandinella*, Pc — *Paramecium caudatum*, Pb — *Paramecium bursaria*, Pt — *Paramecium trichium*, Ut — *Urocentrum turbo*.

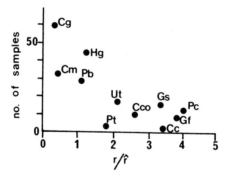

Source: Taylor (1978b), with the permission of Springer-Verlag, Heidelberg.

Isolating populations in the laboratory and testing the r-K selection hypothesis can inevitably lead to anomalies. For example, using K as an index of competitive ability does not take interference competition into account (Gill, 1972), and neglects a whole array of environmental factors which impose an effect on feeding, growth and reproductive physiology of an organism, often over a short period in the natural environment. One factor which must not be overlooked is the very short life-cycle of protozoans compared to metazoans, so that the characteristics of a population can change quickly in terms of numerical density, mean cell volume or size and overall biomass. As Taylor (1978b) very rightly points out, an organism pursuing a persistent or K-'strategy' in an unstable environment, may, if transferred to a stable one, succeed in the role of an r-'strategist'. An organism's position on the r-K continuum is influenced by the abiotic conditions imposed by the environment and the nature of the competition it encounters. Thus Protozoa cannot be recognised as being either r- or K-strategists; their position on the continuum is a function of an array of biotic and abiotic variables.

D. Interactions of Bacterivorous Protozoa with their Food Source

Microorganisms, particularly bacteria, play a role in the flow of carbon, nitrogen and phosphorus in nature, and additionally act as an energy source to a range of microbivorous organisms, including many species of Protozoa. The interactions of the Protozoa with their food source have been considered in a traditional predator—prey context. Early studies on the impact of Protozoa on their bacterial food source suggested that the relationship was not simply one of a secondary consumer exploiting a primary consumer; there seemed to be a stimulation of bacterial growth as a consequence of the presence and/or the grazing activity of the Protozoa. Since bacteria perform an important role in releasing nutrients from dead organic matter, and since their growth appeared to be modified by the presence of Protozoa, it follows that Protozoa may play an indirect role in nutrient recycling. Interest in decomposition processes and the recycling of nutrients in nature has grown in recent years, and evidence is now accumulating to suggest that Protozoa may play a direct, as well as an indirect role in the recycling of some nutrients, particularly phosphorus.

(i) The Impact of Protozoa on their Bacterial Food Source

Much of the early work on this aspect of protozoan ecology was directed towards soil-dwelling species, and in particular the impact on nitrogen-fixing bacteria in the soil. Later, interest turned to the aquatic environment. All these studies indicate that usually the impact of protozoan grazing on bacteria is an increase in bacterial production and a hastening of the oxidation of decomposing organic matter in the soil or in water. Exactly how Protozoa achieve this effect is a matter of some debate. It has been suggested that Protozoa prevent the bacteria from reaching self-limiting numbers, and thus maintain the bacteria in a prolonged state of high metabolic activity or in a state of physiological youth (Imhoff and Fair, 1961). Other explanations include the secretion of growth-promoting substances by Protozoa (Hervey and Greaves, 1941; Strǎsbrabová-Prokešová and Legner, 1966; Nikoljuk, 1969).

The level of nitrogen fixation by *Azotobacter* in the presence and absence of protozoan grazing has been widely investigated. It is a comparatively easy model to monitor in the laboratory. One of the earliest observations was made in 1909 by Russell and Hutchinson, who observed an increase in ammonia production when Protozoa were present in soil. Later' Nasir (1923) and Cutler and Bal (1926) showed that increased nitrogen fixation by *Azotobacter* occurred when the bacterium was grazed by ciliates or amoebae. In the presence of *Azotobacter* and other bacteria, nitrogen fixation in 50 ml of medium containing 0.25 g of mannitol amounted to 0.81-1.10 mg over 15 days, but in the presence of *Colpidium*, nitrogen fixed was 2.42 mg. Where *Hartmannella* grazed the bacteria, 1.69-2.21 mg of nitrogen was fixed under the same conditions in the 15-day period (Cutler and Bal, 1926). The ciliates and amoebae had the effect of reducing bacterial numbers, while enhancing nitrogen fixation by the remaining bacteria. The phenomenon was attributed to a greater efficiency in the use of the substrate where bacterial density was reduced by protozoan feeding activity.

Hervey and Greaves (1941) also found that the quantity of nitrogen fixed in the presence of a ciliate, in this case *Colpoda*, was greater than when *Azotobacter* was ungrazed. They considered a series of possible mechanisms which may have been responsible for the effect they observed. One possibility was the view that bacteria were maintained in physiological youth by protozoan grazing. Alternatively the Protozoa and bacteria had a type of mutualistic relationship, the bacteria serving as food, while the Protozoa excreted ammonia which maintained the alkalinity of the medium by neutralising the organic acids present. Neither theory sufficiently explained their results when tested, so they looked for other causes. They discovered that the addition of dead Protozoa still increased nitrogen fixation, which led them to the conclusion that Protozoa must secrete a growth-promoting substance. They suggested that it was probably a protein of an enzymatic nature, acting either as an absorbing agent rendering certain ions more accessible, or carrying vitamins or vitamin-like substances.

The concept of protozoans secreting a bacterial growth-promoting substance has been considered several times since it was first mooted by these authors (Hervey and Greaves, 1941). In aquatic situations ciliates enhanced the multiplication of bacteria in the presence of glucose during the initial 12-24 hours of incubation, during which time insignificant protozoan multiplication occurred. The same effect on bacterial growth was achieved by adding liquid from which ciliates had been removed, suggesting the presence of some stimulatory substance. The stimulatory effect was destroyed by heating to boiling, which possibly indicates the denaturing of a protein (Stráskrabová-Prokešová and Legner, 1966).

Nikoljuk (1969) found that amoebae and ciliates increased nitrogen fixation by bacteria in a range of soil types. Flagellates, however, which also consumed bacteria, had little effect on the level of nitrogen fixation by soil bacteria. The fact that some protozoan groups can stimulate increased nitrogen fixation and some do not to any great extent, may be a function of the varying impact of their grazing on the bacteria, but alternatively may be the result of the ability of some Protozoa to produce growth-promoting substances while others do not possess the capability to do so. Nikaljuk considered this possibility, and found that a growth-promoting substance was apparently produced, and in the case of amoebae was identified as indole-3-acetic acid (IAA) or heteroauxin, which is a plant growth substance. *Azotobacter* did not apparently produce IAA itself, but subsequently Brown and Walker (1970) found small amounts of IAA in aerated liquid cultures of *Azotobacter chroococcum*. However, the fact that the bacterium produces IAA itself does not preclude an enhancement of growth by additional IAA secreted by Protozoa. The production of this substance which stimulates plant growth is not unique to the Plant Kingdom, for human urine is a rich source of indole-3-acetic acid (Thimann, 1979).

The stimulatory effect of Protozoa on nitrogen-fixing bacteria has been shown to be temperature-dependent. Experiments with the ciliate *Colpoda steini* and *Azotobacter chroococcum* at a range of temperatures produced very varied results. At 28°C more nitrogen was fixed by the bacteria on their own, than in the presence of the ciliate, while at 25°C there was no significant difference in the amount of nitrogen fixed in the presence or absence of protozoan grazing. At 15°C, however, cultures with *Colpoda* produced twice as much nitrogen as pure bacterial cultures. At 5°C the ciliate was unable to survive (Darbyshire, 1972). These results suggest that *Colpoda* may be near its thermal limit at 28°C; this is a high temperature and not one experienced for any length of time in temperate soils, although 28°C happened to be the optimum temperature for nitrogen fixation by *Azotobacter*. At 5°C the ciliate did not function at all, which suggests an optimum temperature for *Colpoda* around 15-20°C; at 15°C the ciliate was found to enhance nitrogen fixation. Darbyshire (1972) gave support to the growth substance theory with the proviso that it functioned only at some temperatures. In the soil a wide range of protozoan species occurs, each with different temperature tolerances. Thus at any temperature there will always be some species able to function efficiently and presumably maintain their intimate stimulatory effect on their bacterial food source.

The stimulatory action of Protozoa on bacterial production has recently been considered in the decomposer communities of aquatic environments. Fenchel (1977) showed a reduction in bacterial numbers to 50 per cent of the ungrazed density, when a mixed protozoan fauna was present. A single protozoan species succeeded in reducing bacterial density by 70 per cent. Much of the detritus in freshwater aquatic environments is plant-derived, and it has been suggested that the bacterial density on such detritus is closely related to the free surface area and that it is the availability of surface which may limit bacterial production (Fenchel and Jørgensen, 1977). In the presence of protozoan grazers, 20 per cent or less of the surface area of detritus is covered by bacteria, but in the absence of Protozoa all the available area on the detritus is colonised by bacteria. Another possibility suggested by Fenchel and Jørgensen (1977) is that microturbulence created by the Protozoa may have a stimulatory impact on bacteria.

(ii) The Role of Protozoa in Nutrient Recycling

Nutrient regeneration can be defined as the release of soluble organic and inorganic nutrients, necessary for primary production, from dead organic matter. In terrestrial ecosystems decomposition processes brought about by microbial activity are confined to the soil and litter layers, whereas in aquatic environments they not only occur on and in the sediments of the benthic zone, which are analagous to soils, but in the planktonic communities of the water as well. Most of the earlier work on decomposition processes was focused on the soil habitat, where bacteria are one of the major elements responsible for the process of nutrient cycling. The relatively recent quantitative studies of decomposition and nutrient regeneration in aquatic environments indicate that bacteria are not the only important group of organisms involved. Indeed, much of the phosphorus and nitrogen incorporated into aquatic primary and secondary production may be regenerated by processes other than bacterial action (Johannes, 1968).

In aquatic habitats, many organisms, including zooplankton and

Protozoa play a role as decomposers and nutrient regenerators, because as a consequence of their metabolising the organic molecules in their food, they excrete nutrients. In this way quantities of phosphorus in the organic orthophosphate form are released into the environment (Pomeroy *et al.*, 1963; Satomi and Pomeroy, 1965). Nitrogen is released in ammonia-free amino acids and other compounds (Johannes, 1968). Phosphorus is important in energy and nutrient cycles, and Protozoa have been shown to play an important role in the recycling of this element in marine waters. Johannes (1964) showed that per unit weight dissolved phosphorus excretion rates by marine animals increased with decreasing organism size. Since Protozoa are small and have a correspondingly high metabolic rate, they have a high rate of phosphorus excretion. The time taken to excrete the equivalent of the body content of phosphorus was less than one hour in the ciliate *Euplotes* (Johannes, 1965). This is extremely high when compared with zooplankton, which take 1.5-3 days to release the equivalent of their body content of phosphorus (Pomeroy *et al.*, 1963).

Conflicting results were found in a tundra freshwater habitat by Barsdale *et al.* (1974). While they were able to demonstrate that the transfer of phosphorus from detritus through bacteria to solution was enhanced by the presence of grazers, little of the phosphorus actually passed through the grazers in the process, because the turnover time of bacterial phosphorus due to grazing activity was about 24 hours, whereas the turnover of phosphorus in the bacterial biomass was only an hour or less. They were unable to explain the exact mechanism whereby phosphorus cycling occurred more rapidly when Protozoa were grazing bacteria than when bacteria were ungrazed, but suggested that protozoan removal of bacteria modified bacterial physiology and promoted a more rapid assimilation of the phosphorus contained in detritus and a more rapid circulation of phosphorus. Whatever the cause, conflicting though the evidence is from different types of ecosystem, it is clear that Protozoa play a valuable direct and/or indirect role in the cycling of phosphorus.

Carbon is one of the most important elements cycled in nature, since it forms one of the major components of living cells. Two areas of an organism's physiology are important in the carbon cycle, one being the intrinsic rate of increase of an animal and the second being the rate of respiration, or the fermentation, of carbon. These two aspects determine the relationship between assimilated carbon and loss through respiration. Protozoa, by virtue of their rapid rates of growth and division and high assimilation efficiencies, will sustain a larger population on a given quantity of nutrients and energy than metazoans, and will also attain greater efficiency in the conservation of nutrients and energy. Hence the physiological characteristics of Protozoa confer upon them a valuable role in the cycling of carbon (Stout, 1980). In experimental soil microcosms, where bacteria were grazed by amoebae, which in turn were predated by nematodes, the release of carbon through respiration was significantly higher than in ungrazed bacterial controls (Coleman *et al.*, 1978). These workers found that in an experimental situation where fixed amounts of carbon were available, partitioning of carbon to biomass and respiration varied in relation to the complexity of the food chain and the assimilation efficiencies of the organisms involved.

Carbon is not usually a factor limiting primary production in terrestrial habitats, but may be so in aquatic environments, where the available CO_2 in the water may limit photosynthesis. However, it is normally the availability of nitrogen and phosphorus in any system which imposes constraints on the rate of primary production and bacterial decomposition. Where these essential elements are not available in the dead organic matter substrate in amounts sufficient to sustain bacterial growth, they are assimilated by the bacteria from the surrounding environment, so that in some circumstances the decomposers may be in direct competition with the primary producers for nitrogen and phosphorus. In the soil this phenomenon is known as nutrient immobilisation, because it is a process in which bacteria reduce the quantities of nutrients available to plants (Alexander, 1961). In aquatic environments bacteria can absorb phosphorus and nitrogen from the water and have the capability of storing phosphorus in excess of their immediate needs. It follows that high bacterial densities in the soil or sediments and waters of aquatic ecosystems is not necessarily indicative of a high turnover of nutrients. Studies on phosphorus transformations in artificial soil communities show that bacteria very quickly assimilate and retain most of the labile inorganic phosphorus available. Bacterial phosphorus is released only slowly when bacterial grazers are absent. When in the presence of bacterivorous amoebae or other protozoans, however, phosphorus is rapidly mineralised and returned as inorganic phosphorus to the pool (Cole *et al.*, 1978). Much faster nutrient recycling occurs extrinsic to the organic substrate between bacteria and bacterial grazers, and it is this part of the mineral recycling which controls the rate of decomposition processes and in addition the structure of the decomposer communities. The grazing activities of Protozoa on bacteria, their stimulatory effect on the rate of bacterial production, and their own intrinsically high rates of growth and reproduction all contribute to the valuable role played by these microfauna in the regeneration of nutrients.

E. Predators of Protozoa

Protozoa, along with other elements of the micro- and meiofauna, are subject to predation. There are considerable difficulties in identifying the predators of Protozoa by the usual procedures. Protozoa do not leave easily identifiable remains when consumed by a predator, so the techniques of gut and faecal analysis are of no value. Direct observation, usually in the laboratory, is the only reliable means of determining which predators readily exploit Protozoa as an energy source.

Most of the evidence in respect of predation on Protozoa pertains to aquatic environments. The medium lends itself more readily to laboratory investigations based on observation, whereas similar studies on soil are logistically more difficult. There is no doubt that Protozoa must be ingested by detritivorous soil organisms, since the protozoans will be intimately associated with the dead organic matter and its bacterial flora. The question arises as to whether Protozoa are an essential source of energy to such organisms. Some experiments with the earthworm *Eisenia foetida* indicate that Protozoa are necessary for normal growth and are a valuable part of the normal diet (Miles, 1963). Among the Nematoda there are a variety of trophic types, including carnivores and bacterivores. It has long been suspected that carnivorous nematodes exploit Protozoa as part of their repertoire of prey. Recently this has been confirmed from studies on trophic interactions in experimental soil microcosms, where populations of *Acanthamoeba polyphaga* were reduced by the predation activities of the nematode *Mesodiplogaster lheriteri* (Anderson *et al.*, 1978).

In the aquatic environment crustacean elements in the zooplankton and benthic communities prey on Protozoa. Both filter-feeding Cladocera and Copepoda and raptorial-feeding Copepoda take a variety of Protozoa, and ciliates in particular. There is considerable evidence to show that Protozoa are a significant element in the diet of these small crustaceans. The raptorial benthic-feeding copepod *Acanthocyclops bicuspidatus* has been fed experimentally on algae, detritus and a range of ciliate species. The development times of *A. bicuspidatus* raised on protozoan diets were one-third to one-half the development times on algal or detritus diets (Strachan, 1980). Since the copepods maintain good production on Protozoa when compared to the other food sources available in the benthos, it is not unreasonable to suppose that copepods actively select these unicellular organisms as food in the natural environment.

Ciliates have been estimated as contributing as much as 20 per cent of the plankton biomass of inshore marine plankton communities, but until fairly recently have been overlooked as a potential energy source to other faunal elements in the zooplankton (Berk *et al.*, 1977). Porter (1973) noticed that ciliate numbers were reduced when planktonic copepods were present compared to controls where copepods were absent. Later Berk *et al.* (1977) showed that *Eurytemora*, a calanoid copepod, can significantly reduce small ciliate numbers, and further that the copepods derived nutritional benefit from them.

Ciliated Protozoa also appear to be a significant food source to zooplankton in freshwater environments. The filter-feeding cladoceran *Daphnia magna* extracts and consumes ciliates from the water filtered during feeding. Where large ciliates, such as *Paramecium caudatum*, are preyed on, only some of those entering the feeding current were brought inside the carapace, and into the food groove; most were broken up by the movement of the thoracic appendages, labrum and mandibles and ingested, but with the loss of some material. Fairly high levels of assimilation efficiency, of the order of 65-66 per cent, are achieved by *D. magna* on a ciliate diet (Porter *et al.*, 1979).

Field studies indicate that planktonic Protozoa, especially flagellates and ciliates, are distributed in definite layers in the water column of lakes. In Dalnee Lake (Kamchatcka) maximal numbers in mid-summer occurred around 16-18 metres depth, which more or less coincided with the thermocline (Sorokin and Paveljeva, 1972). The thermocline is the area of interface between the unmixed lower hypolimnion and the upper epilimnion in stratified lakes. The maximal biomass of Protozoa developed in the period of algal death and the simultaneous development of large bacterial populations, at the end of the spring algal bloom. During the latter part of July the protozoan density decreased. Sorokin and Paveljeva (1972) attributed the decrease to predation by planktonic copepods and the rotifer *Asplanchna priodonta*. In July one-third of the primary food resources in the pelagic community of the lake are formed via bacterial mobilisation of dead organic matter. The most important role in the secondary trophic level is performed by Protozoa, which make a significant contribution to overall zooplankton production. The role played by Protozoa and their interactions with other organisms in Dalnee Lake is shown in Figure 5.6.

Figure 5.6: The Role of Protozoa in the Plankton of a Lake Community. The figures show production for 30 days in June. P = production, the numbers in squares are rations and the numbers in circles are non-assimilated food.

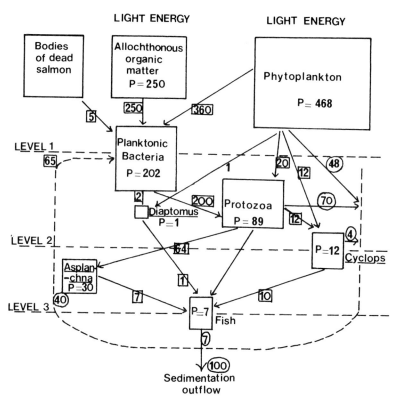

Source: Sorokin and Paveljeva (1972), with the permission of Dr W. Junk, Publishers.

The value of Protozoa as a food source to other organisms in the food chains of soil and aquatic benthic or planktonic communities is an area in which we have only limited information at present. Undoubtedly, as the realisation that Protozoa play a role in decomposition processes and the recycling of nutrients becomes more widely appreciated, their impact further up the food chain as a source of potential energy to secondary and tertiary consumers will also attract the attention of ecologists, in their attempts to elucidate pathways of energy flow and nutrient cycling in nature.

6 ECOLOGY AND ADAPTABILITY

A. Introduction

Despite the fact that Protozoa are ubiquitous, they are frequently overlooked in faunal surveys of ecosystems, but as we have seen in the preceding chapter they play a valuable role in the functioning of many ecosystems. Much of their success in colonising so varied an array of habitats is due to their physiological and behavioural adaptability. Through time various assemblages of species have evolved physiologically to meet the variety of physical, chemical and biological conditions imposed by a range of ecological niches, many of which are inhospitable or hazardous. One of the advantages of a single-celled organisation is that it is often more readily able to undergo evolutionary adaptation and change than a multicellular level of organisation. The widespread ability of Protozoa to encyst, which is a characteristic shared by a range of distinct taxonomic groups in the Protista, has proved invaluable to those species which have successfully moved from aquatic environments to terrestrial and semi-terrestrial habitats.

The position of many natural protozoan populations in the saprovore food web has resulted in their ready colonisation of artifical systems set up by man to deal with waste water and sewage. Sewage treatment plants are essentially simple artificially manipulated saprovore systems aimed at purifying the waste organic materials of urban industrialised society. The protozoan communities of sewage treatment plants, particularly ciliates, perform a valuable role in the production of good quality clean effluents (Curds, 1973, 1975; Curds and Cockburn, 1970a, b).

The small size, rapid growth rates and the ease with which some species can be cultured in the laboratory, together with the tolerance of many Protozoa to polluted conditions, has led to the consideration of these organisms as indicator species, and as species for bioassay in pollution studies. Perhaps more importantly the impact of bacterivore Protozoa on the population dynamics of bacteria and in turn on decomposition processes, suggests a valuable role in polluted environments where they may enhance the decomposition of polluting agents of organic origin.

B. Ecology and Adaptation in the Natural Environment

Protozoa have colonised a wide spectrum of aquatic and terrestrial habitats worldwide from the arctic to the equatorial zones. Many of the studies on soil Protozoa have focused on testate species (Heal, 1962, 1964; Coûteaux, 1975, 1976; Lousier, 1976), which tend to give a biased view of these protozoan communities. In some soils testate Protozoa are a dominant group, as they are in the antarctic soils of Signy Island where ciliates are few and naked amoebae totally absent. Here testates and flagellates constitute the major components of the soil Protozoa (Smith, 1973). In some climates, however, ciliates, flagellates and amoebae may be abundant (Stout, 1962; Rogerson and Berger, 1981b). In mineral and beech litter soils Stout (1962) found ciliates to be more numerous than rhizopods, and interestingly the community tended to be dominated by only a few species – in mineral soils by about five species and in beech soils by about ten. Rogerson and Berger (1981b) found amoebae and flagellates to be the most numerous in garden soil, followed by ciliates, with one recurring helizoan species and an occasional testate amoeba during the period from November to April.

Soil species face variations in the water content of their environment; desiccation is particularly pronounced in the summer months in some climates, and in other soils water becomes locked up in ice during the winter months. Although we talk of terrestrial or soil-dwelling Protozoa, they are not strictly terrestrial because their normal function depends on the presence of the surface films of water around soil particles, or on vegetation such as *Sphagnum*, in which they live. Soil Protozoa are essentially aquatic organisms living in a hazardous environment, particularly where climatic extremes prevail. Physiological adaptation is an essential prerequisite to successful population function among soil Protozoa. Where desiccation of the upper soil levels occurs, the Protozoa can migrate downwards to a limited extent providing there is sufficient oxygen and water, and the soil structure allows it, or alternatively they can encyst. Similarly encystment can be resorted to when soil waters freeze. The evidence suggests that encystment is a readily available refuge when environmental conditions become adverse to the continuance of normal physiological functioning. The process of encystment is described in detail in the following section.

Semi-terrestrial salt marsh ecosystems are usually extremely productive. Webb (1956) found a varied protozoan fauna present in the saltings of the Dee estuary in Cheshire, with ciliates occurring as the most abundant group. The Protozoa of the mudflats, however, showed a less varied community. She was unable to find any correlation between the most common ciliates and gross fluctuations in temperature, salinity or hydrogen ion concentration or in relation to seasonal or tidal changes, which suggests that species commonly found in this habitat are physiologically robust and well adapted to coping with the varying conditions encountered in a salt marsh. Lee and Muller (1973) found that Foraminifera of a salt marsh were highly productive, indeed as productive as other groups, including nematodes. The high productivity of foraminiferans in this ecosystem indicates an important role in the overall energy flow of the community and the cycling of nutrients.

In aquatic ecosystems there are two habitats available to Protozoa, the pelagic and benthic zones. The benthic zone is usually the richest in terms of organic matter and microflora, hence it is usually more heavily populated by protozoans than the open waters. However, flagellates and ciliates commonly occur in the plankton and have been reported as a substantial part of the pelagic community in one freshwater lake (Sorokin and Paveljeva, 1972). In the North Atlantic amoebae have been found to be common in the subsurface of the planktonic zone (Davis et al., 1978). Further detailed studies of pelagic protozoan communities and population dynamics are sorely needed. The indications are that in some aquatic environments planktonic Protozoa may be a significant element in the overall planktonic food web.

Studies devoted to the benthic community as a whole, or focusing solely on the protozoan element in the benthos, have been carried out in aquatic habitats since the beginning of the century. Birge and Juday (1911) noted active representatives of a number of protozoan groups under anaerobic conditions in the hypolimnion of Lake Mendota. More detailed studies of freshwater benthic protozoan communities followed (e.g. Wang, 1928; Moore, 1939; Cole, 1955; Webb, 1961; Goulder, 1971, 1974) and later intensive studies of the benthos of marine and brackish waters appeared (e.g. Fenchel and Jansson, 1966; Muus, 1967; Fenchel, 1967, 1968).

In the majority of lakes, except those which are very shallow, stratification of the waters occurs in a yearly cycle. The result is that for parts of the year the benthic zone is subject to low temperatures and depleted oxygen in temperate lakes. The more eutrophic the lake the greater the depletion of oxygen, so that in highly eutrophic lakes complete deoxygenation of the profundus occurs. Such conditions are not conducive to normal physiological function in obligative aerobic organisms, and consequently changes occur in the abundance and species composition of the benthic protozoan community. Webb (1961) recorded 128 species of Protozoa in a small eutrophic lake, Esthwaite Water, in the English Lake District. The abundance of Protozoa, the majority of which were ciliates, was related to the availability of oxygen in the benthic zone. Later surveys on the same lake (Goulder, 1974) showed that large ciliate species such as *Frontonia leucas*, *Loxodes magnus*, *L. striatus*, *Spirostomum minus*, *S. teres*, *Stentor coeruleus* and *S. polymorphus* generally reached their maxima when the bottom of the lake was oxygenated, but the majority disappeared in the summer months when anoxia prevailed. Other species, which are clearly anaerobic or facultatively so, such as *Caenomorpha medusula*, reached their maxima under deoxygenated conditions. Faced with the adverse summer regimen the normal aerobic protozoan fauna has two avenues of avoidance available — either migration upwards into the water column or laterally to the benthic littoral zone into oxygenated conditions, or encystment.

Detailed studies of the benthic zone and the overlying water column have revealed seasonal migration patterns by many of the species commonly found in the fine silty muds of the lake bottom, when oxygen becomes limited (Finlay, 1981; Bark, 1981). Dense populations of the large ciliate species listed above, of the order of $3,633 \pm 944$ cm^{-2}, are typical from October to May. *Loxodes* sp. and *Spirostomum* sp. dominate the community (Finlay, 1981). During summer stratification most of these ciliates migrate upwards into the overlying water, the most densely populated zone coinciding fairly closely with the 0.3-1.0 mg l^{-1} oxygen isopleths. The vertical migrations involve considerable distances for such small organisms; at certain times the majority of protozoans are some 8-9 metres from the sediment surface. While some species migrate, others such as *Paramecium* disappear completely. A group of small ciliates, less than 150 μm in length, remain on the lake sediments developing large populations. These include *Metopus*, *Caenomorpha medusula*, *Brachionella spiralis* and *Saprodinium dentatum* (Bark, 1981). Thus the benthic habitat is vacated by aerobic forms and a community of sulphide ciliates, similar to those in marine communities described by Fenchel *et al.* (1977), develops on the sediment surface and in the overlying waters (Finlay, 1981). A few of the species which undertake seasonal migrations are also found in low densities on the sediments during anoxia, which has led to the supposition that they may be facultative anaerobes, able to switch from one form of metabolism to another in response to prevalent environmental conditions. The sediments of freshwater lakes are usually fine mud or silt so that penetration to any depth by Protozoa is impossible. Most species occur in the upper centimetre, although some individuals do penetrate down to 3-4 cm in the sediments (Goulder, 1971; Bryant and Laybourn, 1972/3). In stratified lakes the anoxic community is restricted to the sediment surface and its occurrence is seasonally determined by the pattern of stratification.

The migratory behavioural patterns of the normally benthic Protozoa of stratified lakes are particularly interesting. Unlike soil-dwelling species in which encystment is the main means of coping with adverse environmental regimes, freshwater ciliates are able to adopt a behavioural means of avoidance. It is probable that encystment is a less common occurrence in the aquatic environment than in the soil habitat, although undoubtedly species must resort to encystment if they are unable to migrate, respire anaerobically or if their temperature or pH limits are exceeded. As an adaptation to soil colonisation, soil-dwelling protozoans have come to employ encystment more readily, encysting and excysting over short periods as conditions temporarily deteriorate and improve (Rogerson and Berger, 1981b).

In the marine ecosystem the benthic communities of the profundal areas are still largely a mystery, although foraminiferans have been found at depths up to 4,000 metres (Coull *et al.*, 1977). It is the littoral and sublittoral zones which for obvious reasons have been studied in detail. Fenchel (1969) tentatively classified the microfaunal benthic communities of the marine inshore environment into estuarine and sand microbiocenosis, the sublittoral and microbiocenosis and the sulphuretum. The estuarine sand microbiocenosis comprises the microfaunal communities of sandy sediments in fjords, lagoons and sheltered shallow bays. These areas often have high levels of organic material resulting in reduced conditions near to the sediment surface. The sublittoral sand microbiocenosis is found in fine to medium sands possessing a low organic matter content, which are permanently submerged. The sulphureta are biotopes dominated by the sulphur cycle.

The estuarine sand biocenosis has a rich ciliate fauna which is largely confined to the surface oxygenated sediments. A wide spectrum of trophic types are found including species exploiting diatoms, blue-green algae, bacteria, other microfauna and purple and white sulphur bacteria near the sediment surface. The lower reduced sediments contain ciliates exploiting sulphur bacteria under anaerobic conditions. The dominant process in marine sediments is sulphate reduction; the sulphide produced may be bound to iron in the anoxic zones producing a characteristic black colouration. The specialised sulphur ciliates living in these zones are members of the Trichostomatida, Heterotrichida and Odontostomatida. The latter order contains only anaerobic species which lack cytochrome oxidase and mitochondria (Fenchel, 1978). The sublittoral sand microbiocenosis harbours a large number of characteristic ciliates, many of which feed exclusively on diatoms. Bacterivores and predators are also found in this group. Sulphureta, which are common along the shores of many fjords, lagoons and inner bay waters, represent parts of the coastal and marine ecosystem dominated by the sulphur cycle. Thriving populations of ciliates have established themselves in this habitat; Fenchel (1969) recorded some 45 species of ciliate among the sulphur bacteria in the sulpuretum of Nivå Bay.

C. Encystment

Encystment is a widespread phenomenon among taxonomically diverse protozoan groups. Undoubtedly the ability to encyst plays a valuable role in the life strategies of Protozoa, and since it is a characteristic so widely shared it probably developed early in their evolutionary history. The cyst allows a species to withstand adverse conditions in a state which is in some respects analogous to the conditions of hibernation and diapause in higher animals. Additionally the cyst may act as a dispersal stage for many species, particularly those inhabiting terrestrial and transient aquatic environments. A significant number of Protozoa are dispersed in the air or externally and internally by animals. Cysts have been observed in aerial and animal samples, which when cultured have given rise to viable organisms (Corliss and Esser, 1974). The resting cyst plays a valuable role in the life-cycle of those species having the capability of entering such a physiological quiescent state.

During the process of encystment, a series of very radical changes occurs in the cell. There is considerable dedifferentiation, both morphologically and physiologically. The most obvious morphological changes are the loss of the organellar systems of cilia and flagella in most ciliates and flagellates, the eventual loss of the contractile vacuoles and food vacuoles and changes in the structure and distribution of ultrastructural components in the cell.

The nuclei and nucleoli undergo marked changes associated with considerable activity, but the changes appear variable among Protozoa. In *Stylonychia mytilus* the macronuclei fuse to form a single macronucleus, and the chromatin becomes organised into discrete round or oval-shaped bodies. Two types of nucleoli develop in the macronucleus during encystment: a centrally-placed larger nucleolus and a smaller peripherally-located diffuse nucleolus which disappears in the encysted cell (Walker *et al.*, 1975). Fusion of macronuclei also occurs in *Oxytricha*, but here the nucleoli become homogeneous (Grimes, 1973). In both cases the micronuclei resemble those in the vegetative form. The macronucleus of *Blepharisma stoltei* undergoes shortening, adopting a horse-shoe shape and in the later stages of encystment develops a vacuole (Repak, 1968). In *Acanthamoeba* sp. buds form from the nucleolus, which apparently do not contain nucleolar material. The buds are later incorporated into autolysosomes (Griffiths, 1970). A recent study on *Acanthamoeba astronyxis* found that protrusions developed from the nucleoli during encystment which appeared to pinch off as nucleolus-like bodies. One or two of these bodies per nucleus were observed (Lasman, 1982). The exact reasons for all of these variable transformations are unknown.

Mitrochondria also undergo changes in the structure and distribution in the cell. In *Stylonychia mytilus* the mitochondria form tightly packed aggregates which late in encystment form incomplete bands peripherally situated in the cyst. The organelles appear irregular in shape but Walker *et al.* (1975) attributed this appearance to the close packing. Grimes (1973) also noted that mitochondria aggregated in bands in *Oxytricha*; here however they were situated around, but distinct from, the central macronucleus. Changes in mitochondria of *Acanthamoeba* have been reported by Vickerman (1960), who found intra-cristal inclusions. During encystment in *Acanthamoeba astronyxis* circles of rough endoplasmic reticulum appear enclosing protoplasmic inclusions, especially mitochondria. Simultaneously the mitochondria elongate and constrict (Lasman, 1982). Many of these changes are probably associated with changes in the secretory and synthetic activities of the cell during the encystment process. Autolysosomes have been reported during encystment and are prominent throughout the process (de Duve and Wattiaux, 1966; Bowers and Korn, 1969; Holt, 1972; Grimes, 1973). These vacuoles contain various cellular organelles and migrate towards the cell surface where they discharge their contents.

The discharged material may become trapped in the outer layer of the cyst wall, the exocyst.

During encystment there are changes in gross cell composition. A reduction in cell volume occurs, which can be considerable. In *Colpoda steini*, for example, a 60 per cent reduction in cell volume associated with a substantial decrease in cell water content has been reported (Tibbs and Marshall, 1970). As one might expect, contractile vacuole activity is considerable during encystment, increasing in the later stages to achieve cytoplasmic shrinkage. Total cell protein decreases in *Colpoda steini* to 30 per cent of the original amount, and of the remaining protein some 15 per cent is locked up in the cyst wall (Tibbs and Marshall, 1970). Similar reductions in both volume and protein content occur in *Acanthamoeba* where a very large proportion of cell protein, around 33 per cent, is incorporated into the cyst wall (Neff *et al.*, 1964; Neff and Neff, 1969; Rudick and Weisman, 1973). The change in protein content can be followed through growth deceleration and the encystment process. As the organism passes from the exponentially growing phase to the stationary phase and the advent of starvation, which induces encystment in a number of species including *Acanthamoeba* sp., there is a doubling in cell protein levels, followed by a progressive decline during encystment. During the transition to the stationary phase there is extensive proliferation of cytoplasmic membranes, which are subsequently broken down as encystment progresses (Pauls and Thompson, 1978).

The cyst wall is composed of a number of layers which usually range from two to four depending on the species. Indeed there appears to be considerable variation in the structure of the cyst wall among Protozoa. Walker *et al.* (1975) described four layers in the cyst wall of *Stylonychia*: an outer exocyst which appears either amorphous or lamellar; a mesocyst which is fibrous with a lamellar configuration; an amorphous endocyst; and a layer which envelops the cytoplasm composed of compacted membranous material thrown into folds. In *Oxytricha fallax* the exocyst, mesocyst and endocyst are similar to those of *Stylonychia*, but the innermost layer is composed of granular or coarse lamellar material (Grimes, 1973). *Nassula ornata* has two membranes in its cyst wall, a faceted lamellate exocyst and a thin endocyst (Beers, 1966). The exocyst of *Acanthamoeba castellanii* is described as having two layers, an outer amorphous layer overlying a fibrillar layer which may contain some amorphous material, cell debris and glycogen granules (Bowers and Korn, 1969) – Figure 6.1. There may, however, be some variation in the number of layers described by researchers. The exocyst

Figure 6.1: A Cross-section of an Encysted Amoeba. AC — cell debris derived from autolysosomes in the cell wall, En — endocyst, Ex — exocyst, O — operculum over ostioles numbered from 1-6, N — nucleus, n — nucleolus, M — mitochrondria. Scale line equivalent to 1 μ.

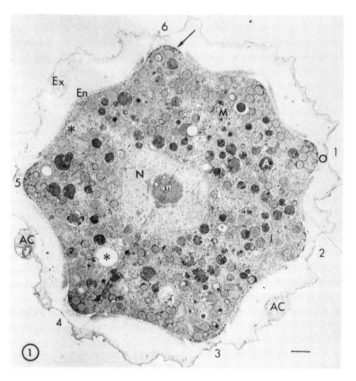

Source: Bowers and Korn (1969). Reproduced from the *Journal of Cell Biology*, 1969, *41*, by copyright permission of the Rockefeller University Press.

described by Bowers and Korn (1969) could be designated as two layers. The endocysts of several species of amoebae, including *Acanthamoeba*, have been shown to contain a cellulose component (Neff *et al.*, 1964; Rastogi *et al.*, 1971). The proportion of cellulose, protein, lipids and carbohydrates in the cyst wall varies from one species of protozoan to another.

While encysted, a protozoan is in a quiescent physiological state. There is no growth or reproduction and the maintenance of living material is continued at a fraction of the respiratory energy expenditure necessary in the vegetative form. Obviously the encystment process itself requires increased metabolic activity which is reflected in increased oxygen uptake, but at the termination of the encystment process the respiratory rate of the cysts in *Hartmanella* was not measurable with Warburg respirometry (Griffiths and Hughes, 1969). Even with the more sensitive cartesian diver microrespirometer the respiratory function of protozoan cysts is not measurable (Laybourn, 1976c). A conservation of energy during the encysted phase, which in extreme cases can last several years, is essential if an organism is to have energy in hand for the energetically expensive process of excystment and commencement of normal locomotion, feeding and growth.

The factors responsible for inducing encystment are probably far more complex than may be apparent from laboratory studies, where often one factor such as depletion of food supply as in *Acanthamoeba castellanii, Oxytricha fallax, Nassula ornata, Blepharisma stoltei* and *Colpoda steini* (Hashimoto, 1962; Repak, 1968; Tibbs and Marshall, 1970; Paul and Thompson, 1978), or high temperature as in *Nassula ornata* collected from low temperature conditions (Raikov, 1962), induces encystment. Indeed, the fact that different strains of one species, *Nassula ornata*, have been found to respond to both lack of food and adverse temperatures by encysting, suggests that both factors operate, or that physiologically distinct strains from one habitat adapted to a particular set of environmental conditions may respond in a different manner to other strains adapted to other regimens. Environmental factors other than temperature have been implicated as inducing encystment and include low oxygen concentration, elevated or reduced pH, increased salt concentration and the accumulation of metabolites (Corliss and Esser, 1974).

We have very little information on why and how encystment occurs under natural conditions, but there are one or two studies which give some indication of the occurrence of encystment in nature and the conditions and factors which may be responsible. Some species of soil and freshwater testate amoebae encyst during the winter months, obviously in response to low temperatures and possibly also reduced food supply, while in the summer months they are active (Schönborn, 1962; Heal, 1964). In the case of zoochlorellae bearing testate species, for example *Hyalosphenia papilio,* encystment in the later part of the year occurs before temperatures fall. In this case it is probably reduced light intensity which induces a reduction in the number of active individuals in the population in *Sphagnum* swards (Heal, 1964). Rogerson and Berger (1981b) observed rapid encystment of the protozoan community of a soil when temperatures declined below $0°C$. Rapid excystment occurred when temperature increased, providing there was sufficient soil moisture. Some intertidal species appear to encyst and excyst in response to a tidal rhythm. *Strombidium oculatum* swims freely at high tide in the water, sinking to the bottom and encysting in intertidal pools at low tide (Fauré-Fremiet, 1948). There appears to be some variability in the ability of the cysts of different species to withstand desiccation. Some freshwater species, for example *Euplotes* and *Didinium*, possess cysts which cannot withstand drying (Beers, 1937; Garnjobst, 1937), whereas the cysts of some soil species can tolerate long periods under dry conditions. Dawson and Hewitt (1931) revived *Colpoda cucullus* after five years of dry storage. The aquisition of resistance to extremes of desiccation in soil species is obviously an evolutionary adaptation to an environment of a more hazardous character than the aquatic ecosystem from which they originated. The extremes of conditions which can be withstood by the cysts of species from different environments is poorly researched, leaving one rather interesting aspect of protozoan ecology and life-history strategies to be elucidated.

The process of excystment requires a stimulus, either replenishment of food supply, suitable temperature or some other factor or combination of factors. The process can be readily induced in the laboratory in some species. *Didinium nasutum* excysts in response to the presence of bacteria. The presence of its prey, *Paramecium*, alone is insufficient (Butzel and Horowitz, 1965). The fact that *Paramecium* does not cause excystment in *Didinium* casts an interesting light on the relationship of this very specialised predator to its prey. *Didinium* cannot encyst without first feeding, so that if it excysts and fails to locate prey, it perishes. However, dense bacteria are likely to attract bacterivorous ciliates including *Paramecium*. *Nassula ornata* rapidly excysts within 2-3.5 hours in the presence of bacteria. Beers (1966) has described the series of events occurring during excystment in *Nassula* in detail. The first sign of cyclosis is the appearance of the contractile vacuole, which swells and having reached a size of about 30 μm, empties very slowly. The process is repeated several times during cyclosis. After 2-2.5 hours a slight elevation develops at the posterior end of the cyst and over the surface of the elevation the exocyst becomes thinner so that its facets are obliterated. It appears that pressure is exerted from within the cyst, probably as a result of the entrance of water, initially into the cytoplasm and later into the contractile vacuole. The contractile vacuole then ceases to evacuate and swells to a diameter of 65 μm. The exocyst ruptures and some of the cytoplasm protrudes abruptly. The form of the rupture is always a slit, which may be as narrow as 15 μm, so that the animal is severely constricted as it passes out, apparently by cytoplasmic streaming even though at this stage the animal is fully ciliated (Figure 6.2).

In other protozoans the cyst wall simply ruptures, presumably as a result of water uptake or in other cases as a result of enzymatic breakdown of the cyst membranes. Often, as in the case of *Didinium nasutum*, a combination of two processes is involved in excystment. In this species a large excystment vacuole forms which causes the rupture and forces off of the exocyst and mesocyst. The vacuole then contracts and a small *Didinium* can be seen swimming inside the endocyst which gradually increases in size and dissolves, thus liberating the enclosed animal (Beers, 1945).

Ultrastructural changes during excystment have been observed in *Oxytricha fallax* and include a complex sequence of micronuclear division. The excysting cell contains dividing, non-dividing and seemingly reabsorbing micronuclei. The function of this micronuclear behaviour is unclear, but in any case, by the time the newly emerged animal is ready to undergo its first division the normal complement of two micronuclei and two macronuclei is established (Grimes, 1973). Dedifferentiation of the ciliature in ciliates occurs during excystment. The first sign in *Oxytricha* is the appearance of a few kinetosomes at the cell surface, which proliferate and become primordia. Then sequentially the cirral primordia, marginal cirral primordia and dorsal bristle primordia develop (Hasimoto, 1963; Grimes, 1973). These structures develop *de novo*, there being no visible markers for the sites of primordia formation. Very shortly after excystment is initiated, the subsurface sheet of microtubules is formed. By the time the animal emerges from the cyst it has a fully formed ciliature.

The ability to encyst has allowed the Protozoa a wider range of habitat colonisation. Soils in which Protozoa are a successful group would in most cases not be a viable environment without the refuge of the encysted state during periods of adversity when desiccation or when freezing of interstitial waters occurs. Terrestrial environments are subject to more rapid and often more severe physical and chemical change than aquatic environments. Encystment for a protozoan is an essential aspect of the life-cycle evolved to withstand periods of adversity. Higher organisms have evolved more sophisticated means of coping with seasonal environmental extremes, by hibernation, diapause or as in the case of many annual invertebrate species, by overwintering the population in the egg or pupal stage. All these mechanisms are designed to cut physiological function to the minimum, so that the organism

Figure 6.2: The Excystment Process in *Nassula*. a: The encysted cell; note the nucleus (n) and the cytopharynx (cp). b: The appearance of the contractile vacuole (cv). c: The rupture of the exocyst. d: The ciliate begins to emerge. e. The endocyst (en) ruptures. f: The ciliate squeezes out of the cyst.

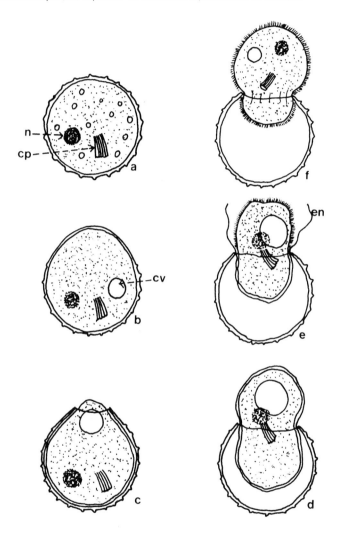

Source: Based on Beers (1966).

passes through an adverse phase when temperatures are low or when food is limited, with the smallest possible expenditure of energy.

D. Protozoa in Sewage Treatment

In addition to colonising a wide array of habitats in natural ecosystems, Protozoa have established themselves in habitats created by man for the treatment of used water. Sewage and used water are treated in a number of ways to produce relatively clean effluents which can be discharged into the environment. In Britain two processes are widely used, the activated-sludge plant and the percolating-filter bed. The former involves the activation of the sludge so that the oxygen incorporated allows bacterial multiplication which leads to oxidation of the organic material in the sewage. The percolating-filter involves trickling the used water, after primary settlement, over a deep bed of small rocks or an artifical medium. The bed has outlets at the bottom through which the purified effluent passes. The medium, usually rocks, is colonised by bacteria which perform the same function as they do in activated-sludge. Activated-sludge plants represent a somewhat unstable environment for colonisation, but nevertheless adaptable elements of the micro- and meiofauna, including Protozoa, Nematoda and Rotifera, have established themsleves in this food-rich environment. Percolating-filters are physically stable habitats and consequently have a greater diversity of animals including copepods, annelid worms, insects and other Metazoa in addition to the groups found in activated-sludge.

Protozoa are abundant in activated-sludge and percolating-filter beds. Densities of 50,000 cells ml^{-1} have been recorded in the mixed liquor of activated-sludge, which represents about 5 per cent of the dry weight of suspended solids in the liquor (Curds, 1973). Achieving quantitative estimates of the densities of organisms in percolating filters is extremely difficult, because the organisms are distributed over the surface of the rocks or plastic media, but observations indicate that Protozoa are abundant.

Essentially the sewage treatment plant is analogous to the polluted natural environment, and the organisms which colonise such habitats must be able to withstand the conditions which are characteristic of pollution. In natural ecosystems severe organic pollution reduces the fauna to all but the most robust species, which often flourish in polluted conditions because of reduced interspecific competition and the rich food supply in terms of dead organic matter and the associated microflora. Protozoa are frequently abundant under such a regimen, hence their successful colonisation of used water processing plants.

Protozoa associated with sewage treatment have been considered by various workers (Ardern and Locket, 1928; Baines *et al.*, 1953) but the most comprehensive study of Protozoa and their role in sewage treatment processes has been carried out by Curds and his co-workers. A survey of 47 percolating filters and 52 activated-sludge plants in the British Isles conducted by Curds and Cockburn (1970a) showed that ciliates were the dominant group, followed by Rhizopoda and Phyto-mastigophorea in descending order of abundance. The Actinopodea were represented by one species, *Actinophrys sol*, which was infrequent in its distribution. In total 67 species of ciliate were identified in activated-sludge plants, while the effluent of percolating-filter beds revealed 53 species. The film scraped from the surface of the medium on the top of the bed showed less species diversity than the effluents, which suggests a succession of species through the depth of the filter bed (Curds, 1975).

In a detailed study of the Protozoa occurring in the activated-sludge process, Curds and Cockburn (1970a, b) found a correlation between the composition of the protozoan community resident in a plant and the quality of the effluent produced. Plants which delivered effluents of high quality with low BOD (Biochemical Oxygen Demand) and low suspended solids, contained a wide variety of ciliated Protozoa in high densities, whereas those plants which produced poor quality effluents contained no ciliates and only a few other Protozoa in low numbers. Laboratory investigations using small-scale activated-sludge plants have proved the important role of Protozoa in waste-water treatment. Under Protozoa-free conditions six replicate plants produced very poor quality turbid effluents, the turbidity being a function of high bacterial numbers and suspended solids. BODs were in the range 53-70 mg l^{-1} and suspended solids ranged between 86 and 118 mg l^{-1}. The addition of cultures of Protozoa produced a very dramatic improvement in the quality of the effluents discharged. After only a few days the BOD dropped to 7-24 mg l^{-1} and suspended solids to 26-34 mg l^{-1}. The improved BOD was attributable to a decrease in viable bacterial numbers from 106-160 $\times$ 10^6 ml^{-1} to 1-9 $\times$ 10^6 ml^{-1} (Curds *et al.*, 1968).

There are several possible explanations for the effect on effluent quality brought about by Protozoa. Bacterivorous ciliates are capable of consuming extremely high numbers of bacteria (see Chapter 2) and voracious grazing on dispersed bacterial growth is probably the major mechanism — indeed Curds believes this to be the case. Some ciliates possess the ability to flocculate suspended particulate matter and bacteria (Watson, 1945; Curds, 1963). This process has also been implicated in the production of good effluents, but Curds (1975) does not consider the process to be of any great import; however, he suggests that amoebae may have the ability to ingest the flocculated bacteria.

The ciliates found in sewage treatment plants can be categorised into three distinct groups: first, free-swimming species which either move in the liquor of activated-sludge or in the surface microbial film of the media in filter beds; secondly, crawling ciliates which crawl on the surface of the floc, and lastly, stalked forms which attach to the floc or media (Curds, 1973). During maturation of activated-sludge there is a succession of protozoan species. Buswell and Long (1923) found that holotrichs were the first to appear, giving way to other forms. Stalked species, such as *Carchesium* and *Vorticella*, appeared after several weeks aeration. More recently Curds (1966) carried out a detailed study of the succession of colonising species in activated-sludge using small-scale experimental plants. Colourless flagellate species, *Oikomonas socialis*, *Peranema trichophorum, Heteronema acus* and *Bodo* were the first to appear. Soon after, ciliates of the free-swimming category appeared, including *Paramecium caudatum*, *Oxytricha fallax*, *Oxytricha ludibunda* and *Uronema nigrans*. These were replaced by crawling hypotrichs, such as *Aspidisca castata* and *Aspidisca lyneus*. When the sludge reached maturity after being aerated for 3-4 weeks peritrichs appeared (Figure 6.3). Curds (1966) considers nutrition to be the major controlling factor in the succession. The pioneering flagellates are either holozoic, saprozoic or saprophytic and all are species normally associated with waters of high organic content. The colonising ciliates then follow successions of bacteria.

E. The Role of Protozoa in Polluted Ecosystems

In industrialised societies pollution is an ever-present problem. The need for rapid and easy tests for indicating the effects of pollution on the flora and fauna and the ecological balance in the environment receiving pollutants, is obvious. Protozoa are near the bottom of the food chain and their small size and rapid reproduction renders the sublethal impact of pollutants more readily discernible than in larger animals with longer, often complex life-cycles. Consequently ecologists

Figure 6.3: The Succession of Ciliates in Developing Activated-sludge.
▲ — flagellates, ● — free-swimming ciliates, ■ — crawling hypotrichs, ○ — peritrichs.

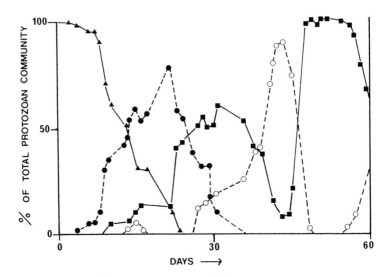

Source: Curds (1966), with the permission of *Oikos*.

have experimented with Protozoa as potential indicator species. An additional aspect is the potential of Protozoa in aiding the breakdown and neutralisation of pollutants. The preceding section dealt with the valuable role of Protozoa in aiding the purification of sewage in artificial habitats, and it follows that as far as organic material is concerned the same function may be performed by protozoans in natural aquatic ecosystems. In oil pollution Protozoa may also play a role in degradation processes.

Aquatic habitats are more prone to pollution than terrestrial environments because water is so widely used in industrial processes, and the effluents of both industry and urban developments are deposited in moving waters, i.e. streams and rivers, estuaries or the sea, for dispersal. Most of these ecosystems can cope with a degree of pollution and usually more severe pollution is reflected in a change in the normal faunal and floral make-up of the community. Severe pollution may create a total anoxic zone which may extend for varying lengths below the effluent outfall.

The value of Protozoa as indicators of organic pollution was recognised early in this century and applied in the saprobity system of water quality classification (Kolkwitz and Marsson, 1908, 1909; Wetzel, 1928; Leibmann, 1936). The saprobity system spans a range from unpolluted to grossly polluted (xenosaprobic – unpolluted, oligosaprobic – scarcely contaminated, β-mesosaprobic – moderately contaminated, α-mesosaprobic – heavily contaminated, polysaprobic – very heavily contaminated). The use of Protozoa as indicators of levels of pollution can be approached in two ways. Either individual species can be investigated as indicators of various degrees of pollution, or species diversity in a community or particular species associations can be considered.

At the species level Sládeček (1964, 1969) produced an indicative value of a species in relation to a degree of saprobity. Thirty-five species of free-moving ciliates were given an index ranging from 1-5 based on their occurrence in nature under particular degrees of saprobity. This approach suggests that some species lend themselves well as indicators, occurring commonly under a particular set of saprobic conditions. Species with a wide distribution through saprobity have obviously less value.

At a more specific level, the effects of isolated chemicals commonly found in industrial effluents have been tested on Protozoa for acute and chronic toxicity evaluation. Apostol (1973) exposed *Paramecium caudatum* to sodium chloride, potassium carbonate, ammonium nitrate, magnesium sulphate, ferrous orthophosphate and lead acetate in concentrations from 1.0 mg 1^{-1} up to 10,000 mg 1^{-1} in acute and chronic tests. The chronic testing involved carrying out experiments in nutrient media to determine any effect on vitality and reproduction, whereas in acute tests the ciliate was simply exposed to the chemicals to evaluate the impact on survival time and any cell damage. Survival time in acute tests decreased with increased toxicant concentration, lead acetate being the most toxic with survival times ranging from 0.1 minute at 10,000 mg 1^{-1} to 500 minutes at 500 mg 1^{-1}. Magnesium sulphate proved the least toxic. Under the chronic testing regime reproductive rate was depressed, usually in relation to increasing toxicant concentration. Bacterial density was not monitored, and there is the possibility that bacterial reproduction was suppressed by a toxicant which would then diminish protozoan food supply, leading in turn to a reduction in reproductive rate in *Paramecium*.

A similar approach, testing the impact of toxicants on an aspect of protozoan physiology as a bioassay technique, was adopted by Slabbert and Morgan (1982). They used the respiratory response of *Tetrahymena pyriformis* to poisons such as ammonia-N, cyanide, PCP, phenol and mercury and elicited a response within 10 minutes to toxicant concentrations in the range 0.5-5.0 mg 1^{-1}. In some cases intoxication caused a decrease in oxygen uptake, in others a marked increase. The latter response was attributed to a state of hypersensitivity indicating that any increase in toxicant concentration would prove lethal. Since the majority of toxicants were detectable at concentrations very much lower than the levels specified for industrial effluents, this technique may be applied successfully for the detection and control of toxic industrial discharges, although at present it is not sufficiently sensitive to detect toxicants at the lower levels permissible in drinking water. However, there is the possibility that other species of ciliates may prove more sensitive to poison and would be more suitable for testing drinking water. The use of Protozoa in bioassay techniques on specific poisons common in industrial effluents is a relatively new area of applied protozoology, but the results indicate that ciliates may prove useful organisms in an important area of ecological monitoring.

In a broader approach, communities of Protozoa may be used for the biological assessment of water quality. Particular associations of organisms are assumed to indicate different degrees of decomposition in organic pollution. In much the same manner that a succession of Protozoa can be observed in the maturation of activated-sludge, successions of associated species can be observed in the process of decomposition in natural waters. Bick (1973) followed the dynamics of protozoan populations in the breakdown of organic matter in a model freshwater ecosystem. The decomposition process takes some time, usually about 7-10 days, to reach the stage where a distinct succession in species and high numbers of ciliates are observed. Successive dominants were *Glaucoma scintillans, Cyclidium, Halteria, Colpidium, Coleps hirtus, Chilodonella cucullulus, Stylonychia putrina, Paramecium caudatum, Litonotus lamella, Acineta* and *Microthorax*. Most of the species are bacterivorous, although one or two carnivores, *Litonotus* and *Acineta*, are present exploiting the other Protozoa as food. These organisms are characteristic of the first three weeks wherein a high intensity of decomposition is established with attendant environmental conditions of oxygen depletion and the presence of ammonia. Bick (1973) terms this a heterotrophic phase. In the later stages when conditions begin to improve an autotrophic phase develops, harnessing the nutrients released by decomposition, when only small numbers of ciliates are present. The large ciliate community is replaced by *Euglena, Chilomonas* and algal species. The succession of species is regulated by biological and environmental factors such as the availability of food, oxygen, the toxic products of decay, competition and predator–prey relations.

The implications of these successions is that protozoan communities clearly play a role in the purification of polluted waters in a manner similar to that which occurs in sewage plants, but also that the community structure is indicative of particular conditions. The real problem comes in correlating the species assemblages present in reasonably accurate terms to the degree of saprobity, especially when one bears in mind that the various stages of saprobity merge into each other. Dresscher and Mark (1976) have devised a simple formula to overcome the problem of the transition from one phase to another and the importance of various groups of Protozoa and algae. The saprobic quotient (x) is calculated as folows:

$$x = \frac{C + _3D - B_{-3}A}{A + B + C + D}$$

The letters indicate the number of forms of each group of organisms. A is ciliates marking polysaprobity, B corresponds to Euglenophyta (Euglenida) indicating α-mesosaprobity, C corresponds to Chlorococcales and Diatomeae indicating β-mesosaprobity, and finally D relates to Peridineae, Chrysophyceae (Chrysomoadida) and Conjugatae indicating oligosaprobity. The limits for the formula are −3 (polysaprobic) to +3 (oligosaprobic). Zero is therefore at the borderline between the condition α-mesosaprobic/β-mesosaprobic and β-mesosaprobic/α-mesosaprobic. This method has the limitation that it cannot be applied when there is a predominance of one species.

In order to monitor an environment quickly for its level of saprobity, it is necessary to sample the water, sediments and macrophytes for the presence of Protozoa. Muddy sediments are time-consuming to analyse and stones and macrophytes are difficult to sample quantitatively. Ideally one requires a rapid means of determining which species are present and their relative abundance. The use of artificial substrates and their colonisation has been investigated by Cairns and his co-workers (Yongue and Cairns, 1978; Cairns et al., 1979), who have shown that the communities which occur on polyurethane foam substrates reflect the natural community. Recently, Henebry and Cairns (1980) have used protozoan communities on polyurethane foam substrates to monitor pollution in a stream subject to enrichment by ammonia, nitrates and phosphates from industrial discharges. Their results suggest that protozoan communities on artifical substrates are an effective means of assessing the degree of pollution, and compare favourably with the macroinvertebrates frequently used as biological indicators. Protozoa on artificial substrates have the advantage of being easy to collect and examine, whereas sampling and collecting macro-invertebrates may be expensive in terms of time.

The evidence to date suggests a valuable role for Protozoa in environmental monitoring of aquatic ecosystems. This is a function of the ecological adaptability of many species and the fact that their physiology responds rapidly to various toxicants, so that sublethal effects are manifested quickly. As bioassay techniques employing Protozoa become more refined there is no doubt that they will become more widely employed in this sphere.

The problem of oil pollution occurs not only in aquatic ecosystems but also in terrestrial ecosystems where oil is extracted, and on salt marshes where estuarine spillages from oil terminals may accumulate by tidal action. There are a variety of bacterial species capable of degrading crude oil. The question arises as to whether there are Protozoa which exploit these bacteria and whether such Protozoa function normally under conditions of oil pollution. Obviously oil-degrading bacteria may be stimulated by protozoan grazing in the same way that bacteria involved in natural decomposition processes are stimulated. Experiments have indicated that crude oil has no adverse effects on freshwater and soil Protozoa, although there is a tendency to slightly larger cells in the presence of oil (Rogerson and Berger, 1981a). The species studied maintained good growth on a diet of oil-degrading bacteria. During feeding some oil is ingested but no deleterious effects are apparent.

High latitude ecosystems are now becoming exploited for oil. Cold temperatures are not conducive to high biological activity and hence to the degradation of oil pollution in such climates. Rogerson and Berger (1981b) investigated the impact of oil pollution on soil in Ontario during the winter months in an attempt to mirror the conditions of more northerly climes. They found that amoebae were the most important group of soil Protozoa, followed in abundance by flagellates and ciliates. Oil had no effect on the protozoan community which exploited the high densities of bacteria associated with the oil. When temperatures dropped to sub-zero the Protozoa encysted, but rapidly excysted when temperatures rose. Indeed during the winter months many species encysted and excysted repeatedly in response to temperature changes although growth and reproduction were minimal during the short active phases. These reoccurring protozoan populations may be expected to have a significant impact on the microflora. Roger-son and Berger suggest that under arctic conditions, where the trophic activity of many terrestrial invertebrates is severely limited during the winter, Protozoa may be extremely important in the functioning of the soil, particularly when subject to pollution.

In aquatic environments oil pollution is normally treated with dispersants. Frequently it is the dispersants rather than the oil which causes death or damage to flora and macrofauna. The impact on Protozoa is poorly researched, although the impact of Corexit 9527 on a range of species has been investigated. The tolerance of each species varied with increasing concentrations. In each case, however, growth rate was constant up to a critical threshold at which point death occurred. Rapid detoxification of cultures occurred in the laboratory, possibly due to evaporation and/or microbial degradation, which suggests that a more rapid recovery from conditions toxic to the most sensitive protozoans would prevail in the field (Rogerson and Berger, 1981c). Since Protozoa can be expected to be relatively unaffected by dispersants in nature it follows that their grazing activities and possible stimulatory impact on oil-degrading bacteria would function even when dispersants are employed.

7 CONCLUSIONS

Despite a single-celled level of organisation the Protozoa are a very successful group of organisms in a variety of ways. By virtue of their adaptable physiology and the widespread ability to encyst they have colonised a wide range of environments, some of which are essentially hazardous to organisms relying on an aqueous medium for normal physiological functioning. Their small size belies the role they perform in food webs, particularly the saprovore food web. A body of evidence is accumulating which implicates some groups of protozoans as stimulatory elements, either directly or indirectly, in the recycling and regeneration of nutrients by the decomposers. The ecological role of Protozoa is multifarious, for not only do they provide a source of energy to predaceous micro-, meio- and macrofauna, but also enhance decomposition processes. In terms of energy transfer between trophic levels their high feeding rates, particularly among bacterivores, and their relatively high assimilation and production efficiencies render them an important component of the ecosystems in which they live, particularly those in which they form a dominant element in the microfaunal communities. There are several well-documented examples of the value and abundance of Protozoa in various environments, but some habitats are yet to be explored in detail, particularly the marine depths. Even in those aquatic and soil habitats which have been investigated quantitatively the information is sketchy and there is a need for more research, particularly with regard to trophic interactions between Protozoa, and between Protozoa and their food sources and predators.

Organellar complexes perform the role undertaken by tissues and organs in higher organisms. The degree of complexity at the single-celled level of organisation is, however, quite startling. Among multicellular animals various types of cells are the building blocks of organs and tissues performing the essential physiological and biochemical functions which sustain life. In many Protozoa, particularly the well-studied ciliates, the microtubule is the essential building element in the construction of a wide variety of cellular systems performing the same fundamental functions as organs and tissues. Membranous structures are frequently also an important component of these cellular complexes. For example, Protozoa lack an alimentary tract, but the food vacuole is a transitory comparable structure. The membranes forming the food vacuoles have been shown in a number of ciliates to form a pool of continually recycling membrane designated specifically for food vacuole formation. Similarly, in the contractile vacuole complex there are indications that membrane may be recycled in those species which lack a permanent vacuole and collecting canals. The level of ultrastructural organisation and biochemical function seen in the locomotory structures of Protozoa, particularly in the ciliates – the evolutionary pinnacle of protozoan development – illustrates very clearly the complexity achievable in a unicellular entity.

The wide diversity in form and physiological function observable in the Protozoa is in part attributable to the taxonomically distant relationships of groups classified in the sub-kingdom. Within discrete taxonomically related groups, however, this diversity is still apparent and is often the result of evolutionary modification imposed by environmental conditions as the Protozoa have colonised different habitats. Within the ciliates, flagellates and sarcodines enormous diversity in form and function occur. A consideration of the feeding physiology or the reproductive physiology of any of these groups illustrates this point.

Our understanding of the intricate details of protozoan structure and function are the result of developments in electron microscopy, and biochemical and physiological techniques. Such techniques are relatively new and are subject to continual refinement, so that the vistas of our understanding of protozoan morphology and physiology are expanding progressively. There is still much to be explored and many intriguing questions are as yet unanswered. For example, we have little information on how pattern formation in developing organellar complexes, especially in ciliary organisation, occurs. Our understanding of the molecular mechanisms involved in the sliding of adjacent doublets in the movement of cilia and flagella is incomplete. The mechanisms involved in co-ordinating movement in sarcodines are unknown. Our knowledge of how membrane is recycled for digestive vacuole formation, and possibly also contractile vacuole formation, in protozoan cells is limited. The present information available on protozoan ecological energetics and interactions between Protozoa and other organisms in the natural environment is only fragmentary.

BIBLIOGRAPHY

Alexander, M. (1961) *Introduction to Soil Microbiology* (Wiley, New York)

Allen, R.D. (1967) 'Fine structure, reconstruction and possible function of components of the cortex of *Tetrahymena pyriformis*.' *Journal of Protozoology, 14*, 553

— (1969) 'The morphogenesis of basal bodies and accessory structures of the cortex of the ciliated protozoan *Tetrahymena pyriformis*.' *Journal of Cell Biology, 40*, 716-33

— (1972) 'Patterns of birefringence in the giant amoeba, *Chaos carolinensis*.' *Experimental Cell Research, 72*, 34-45

— (1974) 'Food vacuole membrane growth with microtubule-associated membrane transport in *Paramecium*.' *Journal of Cell Biology, 63*, 904-22

— (1981) 'Motility.' *Journal of Cell Biology, 91*, 148-55

Allen, R.D. and Fok, A. (1980) 'Membrane recycling and endocytosis in *Paramecium* confirmed by horseradish peroxidase pulse-chase studies.' *Journal of Cell Science, 45*, 131-45

Allen, R.D., Frances, D.W. and Zeh, R. (1971) 'Direct test of the positive pressure gradient theory of pseudopod extension and retraction in amoebae.' *Science, 174*, 1237-40

Allen, R.D. and Staehelin, L.A. (1981) 'Digestive system membranes: freeze-fracture evidence for differentiation and flow in *Paramecium*.' *Journal of Cell Biology, 89*, 9-20

Anderson, O.R. (1980) 'Radiolaria' in M. Levandowsky and S.H. Hutner (eds), *Biochemistry and Physiology of Protozoa* (Academic Press, New York)

Anderson, R.V., Elliott, E.T., McClellan, J.F., Coleman, D.C., Cole, C.V. and Hunt, H.W. (1978) 'Trophic interactions in soils as they affect energy and nutrient dynamics. III. Biotic interactions of bacteria, amoebae and nematodes.' *Microbial Ecology, 4*, 361-71

Andrews, G.A. (1947) 'Temperature effect upon rate of feeding in a folliculinid.' *Physiological Zoology, 20*, 1-4

Apostol, S. (1973) 'A bioassay of toxicity using Protozoa in the study of aquatic environment pollution and its prevention.' *Environmental Research, 6,* 365-72

Ardern, E. and Lockett, W.T. (1928) *Manchester Rivers Department Annual Report,* 41-6

Baines, S., Hawkes, H.A., Hewitt, C.H. and Jenkins, S.H. (1953) 'Protozoa as indicators in activated-sludge treatment.' *Sewage and Industrial Wastes, 25,* 1023-33

Baker, J.R. (1948) 'Status of Protozoa.' *Nature, London, 161,* 548-51

Baldock, B. and Baker, J.H. (1980) 'The occurrence and growth rates of *Polychaos fasciculatum,* a rediscovered amoeba.' *Protistoligica, 16,* 79-83

Baldock, B., Baker, J.H. and Sleigh, M.A. (1980) 'Laboratory growth rates of six species of freshwater Gymnamoebia.' *Oecologia (Berlin), 47,* 156-9

Baldock, B., Rogerson, A. and Berger, J. (1982) 'Further studies on respiratory rates of freshwater amoebae (Rhizopoda;Gymnamoebia).' *Microbial Ecology, 8,* 55-60

Bannister, L.H. and Tatchell, E.C. (1968) 'Contractility and the fibre systems of *Stentor coeruleus.*' *Journal of Cell Science, 3,* 295-308

Bardele, C.F. (1972) 'A microtubule model for ingestion and transport in the suctorian tentacle.' *Zeitschrift für Zellforschung und Mikroskopische Anatomie, 126,* 116-34

— (1974) 'Transport of materials in the suctorian tentacle.' *Society of Experimental Biology Symposium, 28,* 191-208

— (1975) 'The fine structure of the centrohelidan heliozoan *Heterophrys marina.*' *Cell and Tissue Research, 161,* 85-102

— (1977) 'Organization and control of microtubule pattern in centrohelidan Helizoa.' *Journal of Protozoology, 24,* 9-14

Bark, A.W. (1981) 'The temporal and spatial distribution of planktonic and benthic protozoan communities in a small productive lake.' *Hydrobiologia, 85,* 239-55

Barsdale, R.J., Prentki, R.T. and Fenchel, T. (1974) 'Phosphorus cycle of model ecosystems: significance for decomposer food chains and effect of bacterial grazers.' *Oikos, 25,* 239-51

Beers, C.D. (1937) 'The viability of ten-year-old *Didinium* cysts (Infusoria).' *American Naturalist, 71,* 521-4

— (1945) 'The encystment process in the ciliate *Didinium nasutum.*' *Journal of the Elisha Mitchell Scientific Society, 61,* 264-75

— (1966) 'The excystment process in the ciliate *Nassula ornata* Ehrgb.' *Journal of Protozoology, 13,* 79-83

Berger, J. (1979) 'The feeding behavior of *Didinium nasutum* on an atypical prey ciliate *Colpidium campylum.' Transactions of the American Microscopical Society, 98,* 487-94

Berk, S.G., Brownlee, D.C., Heinle, D.R., Kling, H.J. and Colwell, R.R. (1977) 'Ciliates as a food source for marine planktonic copepods.' *Microbial Ecology, 4,* 27-40

Bessen, M., Fay, R.B. and Witman, G.B. (1980) 'Calcium control of waveform in isolated flagellar axonemes of *Chlamydomonas.' Journal of Cell Biology, 86,* 446-55

Bick, H. (1973) 'Population dynamics of Protozoa associated with the decay of organic materials in fresh water.' *American Zoologist, 13,* 149-60

Birge, E.A. and Juday, C. (1911) 'The inland lakes of Wisconsin. The dissolved gases and their biological significance.' *Bulletin of the Wisconsin Geological and Natural History Survey, 22,* 1-16

Bleyman, L.K. (1975) 'Mating type and sexual maturation in *Blepharisma.' Genetics (Suppl.), 80, 14*

Bogdanowicz, A. (1930) 'Uber die konjugation von *Loxodes striatus* (Engelm) Penard und *Loxodes rostrum* (O.F.M. Ehrenb.).' *Zoologischer Anzeiger, 87,* 209-22

Bowers, B. and Korn, E.D. (1969) 'The fine structure of *Acanthamoeba castellanii.* II. Encystment.' *Journal of Cell Biology, 41,* 786-805

Bradfield, J.R.G. (1955) 'Fibre patterns in animal flagella and cilia.' *Symposium of the Society for Experimental Biology, 9,* 306-34

Brown, M.E. and Walker, N. (1970) 'Indolyl-3-acetic acid formation by *Azotobacter chroococcum.' Plant and Soil, 32,* 250-3

Bryant, V.M.T. and Laybourn, J.E.M. (1972/3) 'The vertical distribution of Ciliophora and Nematoda in the sediments of Loch Leven, Kinross.' *Proceedings of the Royal Society of Edinburgh (B), 74, 17,* 265-73

Buswell, A.M. and Long, H.L. (1923) 'Microbiology and theory of activated-sludge.' *Journal of the American Water Works Association, 10,* 309-21

Butzel, H.M. and Horowitz, H. (1965) 'Encystment of *Didinium nasutum.' Journal of Protozoology, 12,* 413-16

Cairns, J., Kuhn, D.L. and Plafkin, J.L. (1979) 'Protozoan colonisation of artificial substrates.' *American Society for the Testing and Materials, Special Technical Publication,* 690

Chapman-Andersen, C. and Prescott, D.M. (1956) 'Studies on pinocytosis in the amoeba *Chaos chaos* and *Amoeba proteus.' Comptes Rendus de Travaux du Laboratoire Carlsberg, 30,* 57-78

Chen, Y.T. (1950) 'Investigations of the biology of *Peranema trichophorum*.' *Quarterly Journal of Microscopical Science, 91,* 279-308

Cole, C.V., Elliott, E.T., Hunt, H.W. and Coleman, D.C. (1978) 'Trophic interactions in soils as they affect energy and nutrient dynamics. V. Phosphorus transformations.' *Microbial Ecology, 4,* 381-7

Cole, G.A. (1955) 'An ecological study of the microbenthic fauna of two Minnesota lakes.' *American Midland Naturalist, 53,* 213-30

Coleman, A.W. (1980) 'Sexuality in colonial green flagellates' in M. Levadnowsky and S.H. Hutner (eds), *Biochemistry and Physiology of Protozoa* (Academic Press, New York)

Coleman, D.C., Anderson, R.V., Cole, C.V., Elliott, E.T., Woods, L. and Campion, M.K. (1978) 'Trophic interactions in soils as they affect energy and nutrient dynamics. IV. Flows of metabolic and biomass carbon.' *Microbial Ecology, 4,* 373-80

Corliss, J.O. (1956) 'On the evolution and systematics of ciliated Protozoa.' *Systematic Zoology, 5,* 68-91, 121-40

—— (1960) 'Comments on the systematics and phylogeny of the Protozoa.' *Systematic Zoology, 8,* 169-90

—— (1961) *The Ciliate Protozoa* (Pergamon Press, Oxford)

—— (1972) 'The ciliate Protozoa and other organisms: some unresolved questions of major phylogenetic significance.' *American Zoologist, 12,* 739-53

—— (1979) *The Ciliated Protozoa,* 2nd edn (Pergamon Press, Oxford and New York)

Corliss, J.O. and Esser, S.C. (1974) 'Comments on the role of the cyst in the life-cycle and survival of free-living Protozoa.' *Transactions of the American Microscopical Society, 93,* 578-93

Coull, B.C., Ellison, R.L., Fleeger, J., Higgins, W., Hope, R.P., Hummon, W.D., Reiger, R.M., Sterrer, W.E., Theil, H. and Tietjen, J.H. (1977) 'Quantitative estimates of the meiofauna from the deep sea off North Carolina, USA,' *Marine Biology, 39,* 233-40

Coûteaux, M.M. (1975) 'Ecologie des Théamoebiens de quelques humis bruts forestiers; l'espèce dans la dynamique de l'equilibre.' *Revue d'Ecologie et de Biologie du Sol, 12,* 421-47

—— (1976) 'Dynamisme de l'equilibre des Thèamoebiens dans quelques sol climactiques.' *Memoires du Muséum Nationale d'Histoire Naturelle, Paris, Ser. A. Zool., 96,* 1-183

Curds, C.R. (1963) 'The flocculation of suspended matter by *Paramecium caudatum*.' *Journal of General Microbiology, 33,* 357-63

—— (1966) 'An ecological study of ciliated Protozoa in activated-sludge.' *Oikos, 15,* 282-9

Curds, C.R. (1973) 'The role of Protozoa in the activated-sludge process.' *American Zoologist, 13,* 161-9

— (1975) 'Protozoa' in C.R. Curds and H.A. Hawkes (eds), *Ecological Aspects of Used-Water Treatment* (Academic Press, London)

— (1977) 'Microbial interactions involving protozoa' in F.A. Skinner and J.M. Shewan (eds), *Aquatic Microbiology* (Academic Press, London)

Curds, C.R. and Cockburn, A. (1968) 'Studies on the growth and feeding of *Tetrahymena pyriformis* in axenic and monoxenic culture.' *Journal of General Microbiology, 54,* 343-58

— (1970a) 'Protozoa in biological sewage-treatment processes. I. A survey of the protozoan fauna of British percolating-filter and and activated-sludge plants.' *Water Research, 4,* 225-36

— (1970b) 'Protozoa in biological sewage treatment processes. II. Protozoa as indicators in the activated-sludge process.' *Water Research, 4,* 237-49

— (1971) 'Continuous monoxenic culture of *Tetrahymena pyriformis*.' *Journal of General Microbiology, 66,* 95-108

Curds, C.R., Cockburn, A. and Vandyke, J.M. (1968) 'An experimental study of the role of ciliated Protozoa in the activated-sludge process.' *Water Pollution Control, 67,* 312-29

Curds, C.R., Gates, M.A. and Roberts, D. McL. (1983) *British and Other Freshwater Ciliated Protozoa. Part III. Ciliophora; Oligohymenophora and Polyhymenophora* (Cambridge University Press, Cambridge)

Curds, C.R. and Vandyke, J.M. (1966) 'The feeding habits and growth rates of some freshwater ciliates found in activated-sludge plants.' *Journal of Applied Ecology, 3,* 127-37

Curry, A. and Butler, R.D. (1982) 'Asexual reproduction in the suctorian *Discophrya collini*.' *Protoplasma, 111,* 195-205

Cutler, D.W. and Bal, D.V. (1926) 'Influence of Protozoa on the process of nitrogen fixation by *Azotobacter chroococcum*.' *Annals of Applied Biology, 13,* 516-34

Dach, H. von, (1942) 'Respiration of a colourless flagellate *Astasia klebsii*.' *Biological Bulletin, 82,* 356-71

Danforth, W.F. (1967) 'Respiratory physiology' in T-T. Chen (ed.), *Research in Protozoology*, Volume I (Pergamon Press, Oxford)

Darbyshire, J.F. (1972) 'Nitrogen fixation by *Azotobacter chroococcum* in the presence of *Colpoda steini*. I. The influence of temperature.' *Soil Biology and Biochemistry, 4,* 359-69

Dass, C.M.S., Kaul, N. and Sapra, G.R. (1982) 'Occurrence of a unique system of intracellular digestive channels in the ciliate *Stylonychia mytilus* Ehrenberg.' *Indian Journal of Experimental Biology, 20*, 244-7

Davis, P., Caron, D. and Sieburth, J. (1978) 'Oceanic amoebae from the North Atlantic: culture, distribution and taxonomy.' *Transactions of the American Microscopical Society, 97*, 73-88

Dawson, J.A. and Hewitt, D.C. (1931) 'The longevity of encysted Colpodas.' *American Naturalist, 65*, 181-6

Dobell, C. (1960) *Antony van Leeuwenhoek and his 'Little Animals'* (Dover Publications Inc., New York)

Dombrowski, H. (1961) 'Methoden und ergebnisse der balneobiologie.' *Ther. Gegen, 100*, 442-9

Dresscher, Th.G.N. and Mark, van der, H. (1976) 'A simplified method for the biological assessment of the quality of fresh and slightly brackish water.' *Hydrobiologia, 48*, 199-201

Dryl, S. (1959) 'The velocity of forward movement in *Paramecium caudatum* in relation to the pH of the medium.' *Bulletin de l'Academie Polonaise des Sciences, Ser. Bio., 9*, 71-4

Duve, C. de and Wattiaux, V. (1966) 'Function of lysosomes.' *Annual Review of Physiology, 28*, 435-92

Edds, K.T. (1975a) 'Motility in *Echinosphaerium nucleofilium*. I. An analysis of particle motions in the axopodia and a direct test of the involvement of the axoneme.' *Journal of Cell Biology, 66*, 145-55

— (1975b) 'Motility of *Echinosphaerium nucleofilium*. II. Cytoplasmic contractility and its molecular basis.' *Journal of Cell Biology, 66*, 156-64

Ettienne, E.M. (1970) 'Control of contractility in *Spirostomum* by dissociated calcium ions.' *Journal of General Physiology, 56*, 168-79

Fauré-Fremiet, E. (1948) 'Le rythme de marée du *Strombidium oculatum*.' *Bulletin Biologique de la France et de la Belgique, 82*, 3-23

Fenchel, T. (1967) 'The ecology of marine microbenthos. I. The quantitative importance of ciliates compared with metazoans in various types of sediments.' *Ophelia, 4*, 121-37

— (1968) 'The ecology of marine microbenthos. III. The reproductive potential of ciliates.' *Ophelia, 5*, 123-36

— (1969) 'The ecology of marine microbenthos. IV. Structure and function of the benthic ecosystem, its chemical and physical factors and the microfauna communities with special reference to ciliated Protozoa.' *Ophelia, 6*, 1-182

— (1975) 'The quantitative importance of benthic microfauna in an arctic tundra pond.' *Hydrobiologia, 46*, 445-65

Fenchel, T. (1977) 'The significance of bacterivorous Protozoa in the microbial community of detrital particles' in J. Cairns (ed.), *Freshwater Microbial Communities,* 2nd edn (Garland Publishing Inc., New York)

— (1978) 'The ecology of micro- and meiobenthos.' *Annual Review of Ecology and Systematics, 9,* 99-121

— (1980a) 'Suspension feeding in ciliated Protozoa: feeding rates and their ecological significance.' *Microbial Ecology, 6,* 13-25

— (1980b) 'Suspension feeding in ciliated Protozoa: functional response and particle size.' *Microbial Ecology, 6,* 1-11

— (1980c) 'Suspension feeding in ciliated Protozoa: structure and function of feeding organelles.' *Archiv für Protistenkunde, 123,* 239-60

— (1980d) 'Relation between particle size selection and clearance in suspension feeding ciliates.' *Limnology and Oceanography, 25,* 733-8

— (1982) 'Ecology of heterotrophic microflagellates. II. Bioenergetics and growth.' *Marine Ecology, Prog. Ser., 8,* 225-31

Fenchel, T. and Jansson, B-O. (1966) 'On the vertical distribution of the microfauna in the sediments of a brackish-water beach.' *Ophelia, 3,* 161-77

Fenchel, T. and Jørgensen, B.B. (1977) 'Detrital food chains of aquatic ecosystems: the role of bacteria.' *Advances in Microbial Ecology, 1,* 1-58

Fenchel, T., Perry, T. and Thane, A. (1977) 'Anaerobiosis and symbiosis with bacteria in free-living ciliates.' *Journal of Protozoology, 24,* 154-63

Fenchel, T. and Small, E.B. (1980) 'Structure and function of the oral cavity and its organelles.' *Transactions of the American Microscopical Society, 99,* 52-60

Finlay, B.J. (1977) 'The dependence of reproductive rate on cell size and temperature in freshwater ciliated Protozoa.' *Oecologia (Berlin), 30,* 75-81

— (1978) 'Community production and respiration by ciliated Protozoa in the benthos of a small eutrophic lake.' *Freshwater Biology, 8,* 327-41

— (1980) 'Temporal and vertical distribution of ciliophoran communities in the benthos of a small eutrophic loch with particular reference to the redox potential.' *Freshwater Biology, 10,* 15-34

— (1981) 'Oxygen availability and seasonal migrations of ciliated Protozoa in a freshwater lake.' *Journal of General Microbiology, 123,* 173-8

Finlay, B.J., Span, A. and Ochsenbein-Gattlen, C. (1983) 'The influence of physiological state on indices of respiration rate in Protozoa.' *Comparative Biochemistry and Physiology, 74A*, 211-19

Finlay, B.J. and Uhlig, G. (1981) 'Calorific and carbon values of marine and freshwater Protozoa.' *Helgoländer Meersunters, 34*, 401-12

Finlay, H.E., Brown, C.A. and Danial, W.A. (1964) 'Electron microscopy of the ectoplasm and infraciliature of *Spirostomum ambiguum*.' *Journal of Protozoology, 11*, 264-80

Fischer-Defoy, D. and Hausmann, K. (1981) 'Microtubules, microfilaments and membranes in phagocytosis: structure and function of the oral apparatus of the ciliate *Climacostomum virens*.' *Differentiation, 20*, 141-51

—— (1982) 'Ultrastructural characteristics of algal digestion by *Climacostomum virens* (Ciliata) (Ehrenberg) Stein.' *Zoomorphology, 100*, 121-30

Fjerdingstad, E.J. (1961) 'The ultrastructure of choancyte collar cell in *Spongilla lacustris* (L).' *Zeitschrift für Zellforschung und Mikroskopische Anatomie, 53*, 645-57

Fok, A.K., Lee, Y. and Allen, R.D. (1982) 'The correlation of digestive vacuole pH and size with the digestive cycle in *Paramecium caudatum*.' *Journal of Protozoology, 29*, 409-14

Garnjobst, L. (1937) 'A comparative study of protoplasmic reorganisation in two hypotrichous ciliates *Stylonethes sterbii* and *Euplotes taylori* with special reference to cystment.' *Archiv für Protistenkunde, 89*, 317-81

Gibbons, I.R. (1963) 'Studies on the protein components of cilia from *Tetrahymena pyriformis*.' *Proceedings of the National Academy of Sciences, USA, 50*, 1002-10

—— (1965) 'Chemical dissection of cilia.' *Archives de Biologie, 76*, 317-52

—— (1966) 'Studies on the adenosine triphosphate activity of 14S and 30S dynein from cilia of *Tetrahymena*.' *Journal of Biological Chemistry, 241*, 5590-6

—— (1981) 'Cilia and flagella of eukaryotes.' *Journal of Cell Biology, 91*, 107-24

Gill, D.E. (1972) 'Density dependence and population regulation in laboratory cultures of *Paramecium*.' *Ecology, 53*, 701-8

Goulder, R. (1971) 'Vertical distribution of some ciliated Protozoa in two freshwater sediments.' *Oikos, 22*, 199-203

—— (1973) 'Observations over 24 hours on the quantity of algae inside grazing ciliated Protozoa.' *Oecologia (Berlin), 13*, 177-82

Goulder, R. (1974) 'The seasonal and spatial distribution of some benthic ciliated Protozoa in Esthwaite Water.' *Freshwater Biology, 4*, 127-47

Grain, J. (1968) 'Les systèmes fibrillaires chez *Stentor igneus* Ehrenberg et *Spirostomum ambiguum* Ehrenberg.' *Protistologica, 4*, 27-36

Grassé, P.P. (1952) *Traite de Zoologie,* Volume I (Fascicle I, Masson et cie, Paris)

Grell, K.G. (1967) 'Sexual reproduction in Protozoa' in T-T. Chen (ed.), *Research in Protozoology*, Volume I (Pergamon Press, Oxford)

— (1974) 'Evolutionary significance of cell division phenomena in Protists.' *Taxon, 23*, 227-8

Griffiths, A.J. (1970) 'Encystment in amoebae.' *Advances in Microbial Physiology, 4*, 105-29

Griffiths, A.J. and Hughes, D.E. (1969) 'The physiology of encystment in *Hartmannella castellanii.*' *Journal of Protozoology, 16*, 93-9

Grimes, G.W. (1973) 'Differentiation during encystment and excystment in *Oxytricha fallax.*' *Journal of Protozoology, 20*, 92-104

Hamilton, R.D. and Preslan, J.E. (1970) 'Observations on the continuous culture of a planktonic phagotrophic protozoa.' *Journal of Experimental Marine Biology and Ecology, 5*, 94-104

Hardin, G. (1943) 'Flocculation of bacteria by protozoa.' *Nature, London, 151*, 642

Harding, J.P. (1937) 'Quantitative studies on the ciliate *Glaucoma*. I. The regulation of the size and the fission rate by the bacterial food supply.' *Journal of Experimental Biology, 14*, 422-30

Hashimoto, K. (1962) 'Relationship between feeding organelles and encystment in *Oxytricha fallax* Stein.' *Journal of Protozoology, 9*, 161-9

— (1963) 'Formation of ciliature in excystment and induced re-encystment of *Oxytricha fallax* Stein.' *Journal of Protozoology, 10*, 156-66

Hausmann, K. and Hausmann, E. (1981) 'Structural studies on *Trichodina pediculus* (Ciliophora: Peritricha). I. The locomotor fringe and the oral apparatus.' *Journal of Ultrastructural Research, 74*, 131-43

Hawes, R.S.J. (1963) 'The emergence of asexuality in Protozoa.' *Quarterly Review of Biology, 38*, 234-42

Heal, O.W. (1962) 'The abundance and micro-distribution of testate amoebae (Rhizopoda: Testacea) in *Sphagnum.*' *Oikos, 13*, 35-47

— (1963) 'Morphological variation in certain Testacea (Protozoa: Rhizopoda).' *Archiv für Protistenkunde, 106*, 351-68

— (1964) 'Observations on the seasonal and spatial distribution of Testacea (Protozoa: Rhizopoda) in *Sphagnum.*' *Journal of Animal*

Ecology, 33, 395-412

Heal, O.W. (1967) 'Quantitative feeding studies on the soil amoebae' in O. Graff and J.E. Satchell (eds), *Progress in Soil Biology* (North Holland Publishing Co., Amsterdam)

Hemmingsen, A.M. (1960) 'Energy metabolism as related to body size and respiratory surfaces, and its evolution.' *Report of the Steno Memorial Hospital and the Nordisk Insulinlaboratorium, Copenhagen, 9,* 1-110

Henebry, M.S. and Cairns, J. (1980) 'Monitoring of stream pollution using protozoan communities on artificial substrates.' *Transactions of the American Microscopical Society, 99,* 151-60

Henk, W.G. and Paulin, J.J. (1977) 'Scanning electron microscopy of budding and metamorphosis in *Discophrya collini* (Root).' *Journal of Protozoology, 24,* 134-9

Hervey, R.J. and Greaves, J.E. (1941) 'Nitrogen fixation by *Azotobacter chroococcum* in the presence of soil Protozoa.' *Soil Science, 51,* 85-100

Hetherington, A. (1932) 'The constant culture of *Stentor coeruleus.*' *Archiv für Protistenkunde, 76,* 118-29

Hibberd, D.J. (1979) 'The structure and phylogenetic significance of the flagellar transition region in the chlorophyll c-containing algae.' *Biosystems, 11,* 243-61

Hiwastashi, K. (1949) 'Studies on the conjugation of *Paramecium caudatum.*' *Sci. Dep. Tohoku Univ., Ser. IV, 18,* 137

—— (1969) 'Paramecium' in C.B. Metz and A. Monray (eds), *Fertilization* (Academic Press, New York)

Holt, P.A. (1972) 'An electron microscope study of the rhadbophorine ciliate *Didinium nasutum* during encystment.' *Transactions of the American Microscopical Society, 91,* 141-68

Holter, H. and Zeuthen, E. (1947) 'Metabolism and reduced weight in starving *Chaos chaos.*' *Comptes Rendus du Travaux du Laboratoire Carlsberg, 26,* 277-96

Holwill, M.E.J. (1966) 'Physical aspects of flagellar movement.' *Physiological Reviews, 46,* 696-785

Holwill, M.E.J. and Silvester, N.R. (1965) 'The thermal dependence of flagellar activity in *Strigomonas oncopelti.*' *Journal of Experimental Biology, 42,* 537-44

Honigberg, B.M., Balamuth, W., Bovee, E.C., Corliss, J.O., Gojdies, M., Hall, R.P., Kudo, R.R., Levine, N.D., Loeblick, A.R., Weiser, J. and Wenrich, D.H. (1964) 'A revised classification of the phylum Protozoa.' *Journal of Protozoology, 11,* 7-20

Hull, R.W. (1954) 'Feeding processes in *Solenophrya micraster* Penard 1914.' *Journal of Protozoology, 1*, 178-82

—— (1961a) 'Studies on suctorian protozoa: the mechanism of prey adherence.' *Journal of Protozoology, 8*, 343-50

—— (1961b) 'Studies on suctorian protozoa: the mechanism of ingestion of prey cytoplasm.' *Journal of Protozoology, 8*, 351-9

Humphreys, W.F. (1979) 'Production and respiration in animal populations.' *Journal of Animal Ecology, 48*, 427-53

Huxley, H.E. and Hanson, J. (1954) 'Changes in the cross-striations of muscle during contraction and stretch and their structural interpretation.' *Nature, London, 173*, 973-6

Hyman, L.H. (1936) 'Observations on Protozoa. I. The impermanence of the contractile vacuole in *Amoeba vespertilis*. II. Structure and mode of food ingestion by *Peranema*.' *Quarterly Journal of Microscopical Science, 79*, 43-56

Imhoff, K. and Fair, G.M. (1961) *Sewage Disposal* (Wiley, New York)

Ivlev, V.S. (1938) 'Sur la transformation de l'energie pendant la croissance des invertèbres.' (in Russian with a French summary) *Byulleten' Moskovskogo Obshestva Ispÿtalelei Prirodÿ, Moskva, 47*, 267-77

Jahn, T.L. and Bovee, E.C. (1964) 'Protoplasmic movements and locomotion of Protozoa' in S.H. Hutner (ed.), *Biochemistry and Physiology of Protozoa* (Academic Press, New York and London)

—— (1967) 'Motile behavior of Protozoa' in T-T. Chen (ed.), *Research in Protozoology*, Vol. I (Pergamon Press, New York)

Jahn, T.L., Bovee, E.C., Fonseca, J.R. and Landman, M. (1964) 'Mechanism of flagellate motion.' *Proceedings of the 10th International Botanical Congress*, Edinburgh, 508

Jahn, T.L., Brown, M. and Winet, H. (1961) 'Secretory activity of oral groove of *Paramecium*.' *Journal of Protozoology, 8 (Suppl.)*, 18

Jahn, T.L., Harman, W.M. and Landman, M. (1963) 'Mechanism of locomotion in flagellates. III. *Peranema, Petalomonas* and *Entosiphon*.' *Journal of Protozoology, 10 (Suppl.)*, 23

James, T.W. and Read, C.P. (1957) 'The effect of incubation temperature on the cell size of *Tetrahymena pyriformis*.' *Experimental Cell Research, 13*, 510-16

Johannes, R.E. (1964) 'Uptake and release of dissolved organic phosphorus by representatives of a coastal marine ecosystem.' *Limnology and Oceanography, 9*, 224-34

—— (1965) 'Influence of marine Protozoa on nutrient regeneration.' *Limnology and Oceanography, 10*, 433-43

Johannes, R.E. (1968) 'Nutrient regeneration in lakes and oceans' in M.R. Droop and E.J. Fergerson-Wood (eds), *Advances in Microbiology of the Sea* (Academic Press, London)

Jones, W.C. (1975) 'The pattern of microtubules in the axonemes of *Gymnosphaera albida* Sassaki, evidence for 13 protofilaments.' *Journal of Cell Science, 18*, 133-55

Karakashian, S.J. (1963) 'Growth of *Paramecium bursaria* as influenced by the presence of algal symbionts.' *Physiological Zoology, 36*, 52-68

Kerkut, G.A. (1960) *Implications of Evolution* (Pergamon Press, Oxford)

Kimball, R.F. (1942) 'The nature of inheritance of mating types in *Euplotes palttia*.' *Genetics, 27*, 265-85

Kimball, R.F., Caspersson, T.O., Svensson, G. and Carlson, L. (1959) 'Quantitative cytological studies on *Paramecium aurelia*.' *Experimental Cell Research, 17*, 160-72

King, C.A., Davies, A.H. and Preston, T.M. (1981) 'Lack of substrate specificity on the speed of amoeboid locomotion in *Naegleria gruberi*.' *Experientia, 37*, 709

Kitching, J.A. (1952) 'Observations on the method of feeding in the suctorian *Podophrya*.' *Journal of Experimental Biology, 62*, 424-37

— (1956) 'Contractile vacuoles of Protozoa.' *Protoplasmologia, 3*, 1-45

— (1967) 'Contractile vacuoles, ionic regulation and excretion' in T-T. Chen (ed.), *Research in Protozoology,* Volume I. (Pergamon Press, New York)

Klekowski, R.Z. and Fischer, Z. (1975) 'Review of studies on ecological energetics of aquatic animals.' *Polskie Archiwum Hydrobiologii, 22*, 345-73

Klekowski, R.Z. and Tumantseva, N.I. (1981) 'Respiratory metabolism in three marine infusoria: *Strombidium* sp., *Tiarina fusus* and *Diophrys* sp.' *Ekologia Polska, 29*, 271-82

Kloetzel, J.A. (1974) 'Feeding in ciliated Protozoa. I. Pharyngeal disks in *Euplotes*: a source of membrane for food vacuole formation?' *Journal of Cell Science, 15*, 379-401

Knight-Jones, E.W. (1954) 'Relations between metachronism and the direction of ciliary beat in Metazoa.' *Quarterly Journal of Microscopical Science, 95*, 503-21

Kolkwitz, R. and Marsson, M. (1908) 'Okologie der pflanzlichen saprobien.' *Bericht der Deutschen Botanischen Gesellschaft, Berlin, 26a*, 505-19

Kolkwitz, R. and Marsson, M. (1909) 'Okologie der tierischen saprobien.' *Internationale Revue der Gesamten Hydrobiologie und Hydrographie, 2,* 126-52

Krogh, A. (1941) *The Comparative Physiology of Respiratory Mechanisms* (University of Pennsylvania Press, Philadelphia)

Kubota, T., Tokoroyama, T., Tsukuda, Y., Koyama, H. and Miyake, A. (1973) 'Isolation and structure determination of blepharismin, a conjugation initiating gamone in the ciliate *Blepharisma,' Science, 179,* 400-2

Kudo, R.R. (1971) *Protozoology,* 5th edn (Charles C. Thomas, Springfield, Illinois)

Lam, R.K. and Frost, B.W. (1976) 'Model of copepod filtering response to changes in size and concentration of food.' *Limnology and Oceanography, 21,* 490-500

Lasman, M. (1982) 'The fine structure of *Acanthamoeba astronyxis,* with special emphasis on encystment.' *Journal of Protozoology, 29,* 458-64

Laybourn, J. (1975a) 'An investigation of the factors influencing mean cell volume in populations of the ciliate *Colpidium campylum.' Journal of Zoology, London, 177,* 171-7

— (1975b) 'Respiratory energy losses in *Stentor coeruleus* Ehrenberg (Ciliophora).' *Oecologia (Berlin), 21,* 273-8

— (1976a) 'Energy consumption and growth in the suctorian *Podophrya fixa* (Protozoa: Suctoria).' *Journal of Zoology, London, 180,* 85-91

— (1976b) 'Energy budgets for *Stentor coeruleus* Ehrenberg (Ciliophora).' *Oecologia (Berlin), 22,* 431-7

— (1976c) 'Respiratory energy losses in *Podophrya fixa* Müller in relation to temperature and nutritional status.' *Journal of General Microbiology, 96,* 203-8

— (1977) 'Respiratory energy losses in the protozoan predator *Didinium nasutum* Müller (Ciliophora).' *Oecologia (Berlin), 27,* 305-9

Laybourn, J. and Finlay, B.J. (1976) 'Respiratory energy losses related to cell weight and temperature in ciliated Protozoa.' *Oecologia (Berlin), 24,* 349-55

Laybourn, J. and Stewart, J.M. (1975) 'Studies on consumption and growth in the ciliate *Colpidium campylum* Stokes.' *Journal of Animal Ecology, 44,* 165-74

Laybourn, J. and Whymant, L. (1980) 'The effect of diet and temperature on reproductive rate in *Arcella vulgaris* Ehrenberg (Sarcodina: Testacida).' *Oecologia (Berlin), 45,* 282-4

Laybourn-Parry, J., Baldock, B. and Kingsmill-Robinson, J.C. (1980) 'Respiratory studies on two small freshwater amoebae.' *Microbial Ecology, 6*, 209-16

Lee, C.C. and Fenchel, T. (1972) 'Studies on ciliates associated with sea ice from Antarctica. II. Temperature responses and tolerances in ciliates from antarctica, temperate and tropical habitats.' *Archiv. für Protistenkunde, 114*, 237-44

Lee, J.J. (1980) 'Nutrition and physiology of the Foraminifera' in M. Levandowsky and S.H. Hutner (eds), *Biochemistry and Physiology of Protozoa* (Academic Press, New York)

Lee, J.J., McEnergy, M., Pierce, S., Freudenthal, H.D. and Muller, W.A. (1966) 'Tracer experiments in feeding littoral Foraminifera.' *Journal of Protozoology, 13*, 659-70

Lee, J.J. and Muller, W.A. (1973) 'Tropho-dynamics and niches of salt marsh Foraminifera.' *American Zoologist, 13*, 215-23

Lee, J.L. (1980) 'Informational energy flow as an aspect of protozoan nutrition.' *Journal of Protozoology, 27*, 5-9

Lee, J.W. (1954) 'The effects of temperature and pH on forward swimming in *Euglena* and *Chilomonas*.' *Proceedings of the Society of Protozoology, 4*, 13-14

— (1956) The effect of pH on the velocity of ciliary movement in *Paramecium*.' *Journal of Protozoology, 3 (Suppl.)*, 9

Leedale, G.F. (1967) *Euglenoid Flagellates* (Prentice-Hall Inc., Englewood Cliffs, New Jersey)

Lehman, J.T. (1976) 'The filter feeder as an optimal forager and the predicted shapes of feeding curves.' *Limnology and Oceanography, 21*, 501-16

Leiner, M., Wohlfeil, M. and Schmidt, D. (1951) 'Das symbiontische bacterium in *Pelomyxa palustris* Greff.' *Zeitschrift für Naturforschung Weisbaden, 6*, 158-70

Levine, N.D., Corliss, J.O., Cox, F.E.G., Deroux, G., Grain, J., Hønigberg, B.M., Leedale, G.F., Loeblich, A.R., Lom, J., Lynn, D., Merinfeld, E.G., Page, F.C., Poljansky, G., Sprague, V., Vavra, J. and Wallace, F.G. (1980) 'A newly revised classification of the Protozoa.' *Journal of Protozoology, 27*, 37-58

Liebmann, H. (1936) 'Auftreten, verhalten und bedeutung der protozoen bei der sekbstrienigung stehenden abwasswers.' *Zeitscrift für Hygiene und Infektionskrankheiten, 118*, 29-63

Lousier, J.D. (1974) 'Effects of experimental soil moisture fluctuations on turnover rates of Testacida.' *Soil Biology and Biochemistry, 6*, 19-26

Lousier, J.D. (1976) 'Testate amoebae (Rhizopoda: Testacea) in some Canadian Rocky Mountain soils.' *Archiv für Protistenkunde, 118,* 191-201

Lövlie, A. (1963) 'Growth in mass and respiration during the cell cycle of *Tetrahymena pyriformis.*' *Comptes Rendus de Travaux du Laboratoire Carlsberg, 33,* 377-413

Luckinbill, L.S. (1979) 'Selection and the r-K continuum in experimental populations of Protozoa.' *American Naturalist, 113,* 427-37

Lynn, D.H. (1975) '*Woodrufia metabolica,* exception to the rule of desmodexy questioned.' *Science, 188,* 1040-1

— (1979) 'Fine structure specializations and evolution of carnivory in *Bresslaua* (Ciliophora: Colpodida).' *Transactions of the American Microscopical Society, 98,* 353-68

— (1981) 'The organization and evolution of microtubular organelles in ciliated Protozoa.' *Biological Reviews, 56,* 243-92

MacArthur, R.H. and Wilson, E.O. (1967) *The Theory of Island Biogeography* (Princeton University Press, Princeton, N.J.)

Machemer, H. (1970) 'Primäre und induzierte Bewegungsstadien bei Osmiunsaürefixierung vorwärtsschwimmender.' *Acta Protozoologica, 7,* 532-5

— (1974) 'Ciliary activity and metachronism in Protozoa' in M.A. Sleigh (ed.), *Cilia and Flagella* (Academic Press, London and New York)

Margulis, L. (1970) *Origin of Eukaryotic Cells* (Yale University Press, New Haven)

— (1974) 'Five-kingdom classification and the origin and evolution of cells' in T. Dobzansky, M.K. Hecht and W.C. Steere (eds), *Evolutionary Biology,* Volume 7 (Plenum Press, New York and London)

— (1981) *Symbiosis in Cell Evolution* (W.H. Freeman, San Francisco)

Mast, S.O. (1906) 'The reactions of *Didinium nasutum* (Stein) with special reference to the feeding habits and functioning trichocysts.' *Biological Bulletin, 16,* 91-118

— (1926) 'Structure, movement, locomotion and stimulation in *Amoeba.*' *Journal of Morphology, 41,* 347-425

— (1947) 'The food vacuole in *Paramecium.*' *Biological Bulletin, 92,* 31-71

Mast, S.O. and Doyle, W.L. (1934) 'Ingestion of fluids by *Amoeba.*' *Protoplasma, 20,* 555-60

Mast, S.O. and Hahnert, W.F. (1935) 'Feeding digestion and starvation in *Amoeba proteus* Leidy.' *Physiological Zoology, 8,* 255-72

Mast, S.O. and Prosser, C.L. (1932) 'Effect of temperature, salts and hydrogen-ion on the rupture of the plasmagel sheet, rate of locomotion and gel-sol ratio in *Amoeba proteus.*' *Journal of Cellular* and *Comparative Physiology, 1*, 333-54

Maynard Smith, J. (1978) *The Evolution of Sex* (Cambridge University Press, Cambridge)

McGee-Russell, S.M. and Allen, R.D. (1971) 'Reversible stabilization of labile microtubules in the reticulopodial network of *Allogromia*.' *Advances in Cellular and Molecular Biology, 1*, 153-84

McKanna, J. (1973) 'Membrane recycling: vesiculation of the amoeba contractile vacuole at systole.' *Science, 179*, 88-90

Metalnikow, S. (1912) 'Contributions a l'étude de la digestion intercellulaire chez les protozoaines.' *Archives de Zoologie Experimentale et Générale, Paris, 9, (4)*, 373-499

Miles, H.B. (1963) 'Soil protozoa and earthworm nutrition.' *Soil Science, 95*, 407-9

Miyake, A. (1974) 'Cell interactions in conjugation of ciliates.' *Current Topics in Microbiology and Immunology, 64*, 49-77

— (1978) 'Cell communication and cell union and initiation of meiosis in ciliate conjugation.' *Current Topics in Developmental Biology, 12*, 37-82

Moore, G.M. (1939) 'A limnological investigation of the microscopic benthic fauna of Douglas Lake, Michigan.' *Ecological Monographs, 9*, 537-82

Muller, M. (1967) 'Digestion' in G.W. Kidder (ed.), *Chemical Zoology*, Volume I (Academic Press, New York)

Muus, B. (1967) 'The fauna of Danish estuaries and lagoons.' *Meddelelser fra Danmarks Fiskeni-og Havundersøgelser, 5*, 1-316

Naitoh, Y. and Kaneko, H. (1973) 'Control of ciliary activities by adenosine-triphosphate and divalent cations in triton-extracted models of *Paramecium caudatum*.' *Journal of Experimental Biology, 58*, 657-76

Nasir, S.M. (1923) 'Some preliminary investigations on the relationship of Protozoa to soil fertility with special reference to nitrogen fixation.' *Annals of Applied Biology, 10*, 122-33

Neff, R.J. and Neff, R.H. (1969) 'The biochemistry of amoebic encystment.' *Symposium of the Society for Experimental Biology, 23*, 51-81

Neff, R.J., Ray, S.A., Benton, W.F. and Wilborn, M. (1964) 'Induction of synchronous encystment in *Acanthamoeba* spp.' *Methods in Cell Physiology, 1*, 56-83

Netzel, H. (1975a) 'Die entstenhung der hexagonalen schalenstruktur bei der thekamöbe *Arcella vulgaris* var *multinucleata* (Rhizopoda, Testacea).' *Archiv für Protistenkunde, 117*, 321-57

Netzel, H. (1975b) 'Morphologie und ultrastruktur von *Centropyxis discoides* (Rhizopoda, Testacea).' *Archiv für Protestenkunde, 117*, 369-92

— (1975c) 'Struktur und ultrastruktur von *Arcella vulgaris* var *multinucleata* (Rihizopoda, Testacea).' *Archiv für Protistenkunde, 117*, 219-45

— (1976) 'Die abscheidung der gehäusewand bei *Centropyxis discoides* (Rhizopoda, Testacea).' *Archiv für Protistenkunde, 118*, 53-91

Nikoljuk, V.F. (1969) 'Some aspects of the study of soil Protozoa.' *Acta Protozoologica, 7*, 99-109

Nisbet, B. (1974) 'An ultrastructural study of the feeding apparatus of *Peranema trichophorum*.' *Journal of Protozoology, 21*, 39-48

— (1984) *Nutrition and Feeding Strategies in Protozoa* (Croom Helm, Beckenham, Kent)

Ogden, C.G. and Hedley, R.H. (1980) *An Atlas of Freshwater Testate Amoebae* (Oxford University Press, Oxford)

Pace, D.M. and Kimura, K.K. (1944) 'The effect of temperature on the respiration of *Paramecium caudatum* and *Paramecium aurelia*.' *Journal of Cellular and Comparative Physiology, 24*, 173-83

Pace, D.M. and Lyman, E.D. (1947) 'Oxygen consumption and carbon dioxide elimination in *Tetrahymena gelii*.' *Biological Bulletin, 92*, 210-16

Patterson, D.J. (1976) 'Observations on the contractile vacuole complex of *Blepharisma americanum* Suzuki 1954 (Ciliophora, Heterotrichida).' *Archiv für Protistenkunde, 118*, 235-42

— (1977) 'On the behaviour of contractile vacuoles and associated structures of *Paramecium caudatum* (Ehrbg).' *Protistologica, 13*, 205-12

— (1978) 'Membranous sacs associated with cilia of *Paramecium*.' *Cytobiologie, 17*, 107-11

— (1980) 'Contractile vacuoles and associated structures: their organization and function.' *Biological Reviews, 55*, 1-46

Patterson, D.J. and Hausmann, K. (1981) 'Feeding by *Actinophrys sol* (Protozoa: Heliozoa). I. Light microscopy.' *Microbios, 31*, 39-55

Patterson, D.J. and Sleigh, M.A. (1976) 'Behaviour of the contractile vacuole of *Tetrahymena pyriformis*: a redescription with comments on the terminology.' *Journal of Protozoology, 23*, 410-17

Pauls, K.P. and Thompson, J.E. (1978) 'Growth and differentiation-related enzyme changes in cytoplasmic membranes of *Acanthamoeba castellanii*.' *Journal of Microbiology, 107*, 147-53

Phillipson, J. (1964) 'A minature microbomb calorimeter for small biological samples.' *Oikos, 15*, 130-9

Phillipson, J. (1981) 'Bioenergetic options and phylogeny' in C.R. Townsend and P. Calow (eds), *Physiological Ecology* (Blackwell Scientific Publications, Oxford/Sinauer Associates, Inc., Sunderland, Mass.)

Pianka, E.R. (1970) 'On 'r' and 'K' selection.' *American Naturalist, 104*, 592-7

Pitelka, D.R. (1968) 'Fibrillar systems in Protozoa' in T-T. Chen (ed.), *Research in Protozoology,* Volume II (Pergamon Press, Oxford and New York)

Pollard, T.D. (1981) 'Cytoplasmic contractile proteins.' *Journal of Cell Biology, 91*, 156-65

Pollard, T.D. and Korn, E.D. (1973) '*Acanthamoeba* myosin. I. Isolation from *Acanthamoeba castellanii* of an enzyme similar to muscle myosin.' *Journal of Biological Chemistry, 248*, 4682-90

Pollard, T.D., Stafford, W.F. and Porter, M.E. (1978) 'Characterization of a second myosin from *Acanthamoeba castellanii*.' *Journal of Biological Chemistry, 253*, 4798-808

Pomeroy, L.R., Mathews, H.M. and Min, H.S. (1963) 'Excretion of phosphate and soluble organic phosphorus compounds by zooplankton.' *Limnology and Oceanography, 8*, 50-5

Porter, K.G. (1973) 'Selective grazing and differential digestion of algae by zooplankton.' *Nature, London, 244*, 179-80

Porter, K.G., Pace, M.L. and Battey, J.F. (1979) 'Ciliate protozoans as links in freshwater planktonic foodchains.' *Nature, London, 277*, 563-4

Prescott, D.M. and Stone, G.E. (1967) 'Replication and function of the protozoan nucleus' in T-T. Chen (ed.), *Research in Protozoology*, Volume I (Pergamon Press, Oxford and New York)

Proper, G. and Garver, J.C. (1966) 'Mass culture of the Protozoa *Colpoda steini*.' *Biotechnology and Bioengineering, 8*, 287-96

Prus, T. (1970) 'Calorific value of animals as an element of bioenergetic investigation.' *Polskie Archiwum Hydrobiologii, 17*, 183-99

Raikov, I.B. (1962) 'Der kernapparat von *Nassula ornata* Ehrbg. (Ciliata, Holotricha). Zur frage über den chromosomenaufbau des makronucleus.' *Archiv für Protistenkunde, 105*, 463-88

— (1969) 'The macronucleus of ciliates' in T-T. Chen (ed.), *Research in Protozoology,* Volume III (Pergamon Press, Oxford and New York)

— (1972) 'Nuclear phenomena during conjugation and autogamy in ciliates' in T-T. Chen (ed.), *Research in Protozoology,* Volume IV (Pergamon Press, Oxford and New York)

Randall, J.T. and Jackson, S.F. (1958) 'Fine structure and function in *Stentor polymorphus*.' *Journal of Biophysical and Biochemical Cytology, 4*, 807-30

Rapport, D.J., Berger, J. and Reid, D.B.W. (1972) 'Determination of food preference of *Stentor coeruleus*.' *Biological Bulletin*, *142*, 103-9

Rapport, E.W., Rapport, D.J., Berger, J. and Kupers, V. (1976) 'Induction of conjugation in *Stentor coeruleus*.' *Transactions of the American Microscopical Society*, *95*, 220-4

Rasmussen, L., Buhse, H.E. and Groh, K. (1975) 'Efficiency of filter feeding in two species of *Tetrahymena*.' *Journal of Protozoology*, *22*, 110-11

Rastogi, A.K., Shipstone, A.C. and Agarwala, S.C. (1971) 'Isolation, structure and composition of the cyst wall of *Schizopyrenus russelii*.' *Journal of Protozoology*, *18*, 176-9

Renaud, F.L., Rowe, A.J. and Gibbons, I.R. (1968) 'Some properties of the protein forming the outer fibers of cilia.' *Journal of Cell Biology*, *36*, 79-90

Repak, A.J. (1968) 'Encystment and excystment of the heterotrichous ciliate *Blepharisma stoltei* Isquith.' *Journal of Protozoology*, *15*, 407-12

Rhumbler, L. (1910) 'Die verchiedenartigen nahrungsaufnahmen bei amoeben als folge verschiedener colloidalzustände ihrer oberflächen.' *Archiv für Entwicklungsmechanik der Organismen*, *30*, 194-223

Ricci, N. and Esposito, A. (1981) 'Mating-type specific soluble factors mediating the preconjugant cell interactions in a marine species of *Blepharisma* (Protozoa, Ciliata).' *Monitore Zoologico Italiano*, *15*, 107-15

Rogerson, A. (1978) 'The energetics of *Amoeba proteus* (Leidy)' (PhD thesis. University of Stirling, United Kingdom)

—— (1979) 'Energy content of *Amoeba proteus* and *Tetrahymena pyriformis* (Protozoa).' *Canadian Journal of Zoology*, *57*, 2463-5

—— (1980) 'Generation times and reproductive rates of *Amoeba proteus* (Leidy) as influenced by temperature and food concentration.' *Canadian Journal of Zoology*, *58*, 543-8

—— (1981) 'The ecological energetics of *Amoeba proteus* (Protozoa).' *Hydrobiologia*, *85*, 117-28

Rogerson, A. and Berger, J. (1981a) 'Effects of crude oil and petroleum-degrading microorganisms on the growth of freshwater and soil Protozoa.' *Journal of General Microbiology*, *124*, 53-9

—— (1981b) 'The effects of cold temperatures and crude oil on the abundance and activity of Protozoa in a garden soil.' *Canadian Journal of Zoology*, *59*, 1554-60

Rogerson, A. and Berger, J. (1981c) 'The toxicity of the dispersant Corexit 9527 and oil dispersant mixtures to ciliate protozoans.' *Chemosphere, 10,* 33-9

Roth, L.E., Pihlaja, D.J. and Shigenaka, Y. (1970) 'Microtubules in the heliozoan axopodium. I. The gradion hypothesis of allosterism in structural proteins.' *Journal of Ultrastructural Research, 30,* 7-37

Rudick, V.L. and Weisman, R.A. (1973) 'DNA-dependent RNA polymerase from trophozoits and cysts of *Acanthamoeba castellanii.*' *Biochemica et Biophysica Acta, 299,* 91-102

Rudzinska, M.A. (1951) 'The influence of amount of food on the reproductive rate and longevity of a suctorian (*Tokophrya infusonium*).' *Science, 113,* 10-11

— (1970) 'The mechanism of food intake in *Tokophrya infusonium* and ultrastructural changes in food vacuoles during digestion.' *Journal of Protozoology, 17,* 626-41

Russell, E.Y. and Hutchinson, H.B. (1909) 'The effect of the partial sterilization of soil on the production of plant food.' *Journal of Agricultural Science, Cambridge, 3,* 111-14 ·

Ryley, J.F. (1967) 'Carbohydrates and respiration' in G.W. Kidder (ed.), *Chemical Zoology,* Volume I (Academic Press, New York)

Salt, G.W. (1974) 'Predator and prey densities as controls of the rate of capture by the predator *Didinium nasutum.*' *Ecology, 55,* 434-9

— (1975) 'Changes in the cell volume of *Didinium nasutum* during population increase.' *Journal of Protozoology, 22,* 112-15

Sarojini, R. and Nagabhushanam, R. (1967) 'A comparative study of some free-living ciliate Protozoa.' *Journal of Animal Morphology and Physiology, 14,* 158-61

Satir, P. (1965) 'Studies on cilia. II. Examination of the distal region of the ciliary shaft and the role of the filaments in motility.' *Journal of Cell Biology, 26,* 805-34

— (1968) 'Studies on cilia. III. Further studies on the cilium tip and a "sliding filament" model of ciliary motility.' *Journal of Cell Biology, 39,* 77-94

Satomi, M. and Pomeroy, L.R. (1965) 'Respiration and phosphorus excretion in some marine populations.' *Ecology, 46,* 877-81

Schaeffer, A.A. (1910) 'Selection of food in *Stentor coeruleus* (Ehrbg.).' *Journal of Experimental Zoology, 13,* 75-132

Scherbaum, O.H. and Loefer, J.B. (1964) 'Environmentally induced growth in Protozoa' in S.H. Hutner (ed.), *Biochemistry and Physiology of Protozoa* (Academic Press, New York and London)

Schiemer, F., Duncan, A. and Klekowski, R.Z. (1980) 'A bioenergetic study of a benthic nematode *Plectus palustris* de Man 1880, throughout its life-cycle.' *Oecologia (Berlin), 44*, 205-12

Schönborn, W. (1962) 'Uber *Planktismus* und zyklomorphose bei *Difflugia limnetica* (Levander) Penard.' *Limnologica, 1,* 21-34

— (1977) 'Production studies on Protozoa.' *Oecologia (Berlin), 27,* 171-84

— (1981a) 'Populationsdynamik und produktion der Testaceen (Protozoa: Rhizopoda) in der Saale.' *Zoologische Jahrbücher, Systematik, Okologie und Geographie der Tiere, 108,* 301-13

— (1981b) 'Die ziliatenproduktion eines baches.' *Limnologica, 13,* 203-12

Sevarin, L.N. and Orlovskaja, E.E. (1977) 'Feeding behaviour of unicellular animals. I. The main role of chemoreception in the food choice of carnivorous Protozoa.' *Acta Protozoologica, 16*, 309-32

Shortess, G.S. (1942) 'The relationship between temperature, light and rate of locomotion in *Peranema*.' *Physiological Zoology, 15*, 184-95

Sibly, R.M. (1981) 'Strategies of digestion and defecation' in C.R. Townsend and P. Calow (eds), *Physiological Ecology* (Blackwell Scientific Publications, Oxford/Sinauer Associates, Inc., Sunderland, Mass.)

Sibly, R. and Calow, P. (1982) 'Asexual reproduction in Protozoa and invertebrates.' *Journal of Theoretical Biology, 96*, 401-24

Slabbert, J.L. and Morgan, W.S.G. (1982) 'A bioassay technique using *Tetrahymena pyriformis* for the rapid assessment of toxicants in water.' *Water Research, 16*, 517-23

Sládeček, V. (1964) 'Zur ermittlung des indikationsgewichtes in der biologischen gewasseruntersuchung.' *Archiv für Hydrobiologie, 60,* 241-3

— (1969) 'The indicator value of some free-living ciliates.' *Archiv für Protistenkunde, 3,* 276-8

Sleigh, M.A. (1956) 'Metachronism and frequency of beat in the peristomial cilia of *Stentor*.' *Journal of Experimental Biology, 33,* 15-38

— (1962) *The Biology of Protozoa* (Pergamon Press, New York)

— (1974) 'Patterns of movement of cilia and flagella' in M.A. Sleigh (ed.), *Cilia and Flagella* (Academic Press, London and New York)

— (1979) 'Radiation of the eukaryotic Protista' in M.R. House (ed.), *The Origin of Major Invertebrate Groups* (Academic Press, London and New York)

Sleigh, M.A. and Barlow, D. (1976) 'Collection of food by *Vorticella*.' *Transactions of the American Microscopical Society, 95*, 482-6

Slobodkin, L.B. and Richman, S. (1961) 'Calories/gm in species of animals.' *Nature, London, 191*, 299

Smith, H.S. (1973) 'The Signy Island terrestrial reference sites. II. Protozoa.' *Bulletin of the British Antarctic Survey, 33, 34*, 83-7

Sonneborn, T.M. (1957) 'Breeding systems, reproductive methods and species problems in Protozoa' in E. Mayr (ed.), *The Species Problem* (American Association for the Advancement of Science, Washington)

Sorokin, Yu. I. and Paveljeva, E.B. (1972) 'On the quantitative characteristics of the pelagic ecosystems of Dalnee Lake (Kamchatka).' *Hydrobiologia, 40*, 519-52

Spoon, D.M., Chapman, G.B., Cheng, R.S. and Zane, S.F. (1976) 'Observations on the behavior and feeding mechanisms of the suctorian *Heliophrya erhardi* (Rieder) Matthes preying on *Paramecium.*' *Transactions of the American Microscopical Society, 95*, 443-62

Stephens, R.E. (1974) 'Enzymatic and structural proteins of the axoneme' in M.A. Sleigh (ed.), *Cilia and Flagella* (Academic Press, London and New York)

Stout, J.D. (1962) 'An estimation of microfaunal populations in soils and forest litter.' *Journal of Soil Science, 13*, 314-20

— (1980) 'The role of Protozoa in nutrient cycling and energy flow.' *Advances in Microbial Ecology, 4*, 1-50

Strachan, I.M. (1980) 'Feeding and population ecology of *Acanthocyclops bicuspidatus* (*sensu stricta*) (Claus) in Esthwaite Water, Cumbria' (PhD thesis, University of Lancaster, UK)

Straškraková-Prokešová, V. and Legner, M. (1966) 'Interactions between bacteria and Protozoa during glucose oxidation in water.' *Internationale Revue der gesamten Hydrobiologie und Hydrographie, 51*, 279-93

Stump, A.B. (1935) 'Observations on the feeding of *Difflugia, Pontigulasia* and *Lesquereusia.*' *Biological Bulletin, 69*, 136-42

Summers, K.E. and Gibbons, I.R. (1973) 'Effects of trypsin digestion on flagellar structures and their relationship to motility.' *Journal of Cell Biology, 58*, 618-29

Takahashi, M. and Tonomura, Y. (1978) 'Binding of 30S dynein with the B-tubule of the outer doublet of the axonemes from *Tetrahymena pyriformis* and adenosine triphosphate-induced dissociation of the complex.' *Journal of Biochemistry, Tokyo, 84*, 1339-55

Tansey, M.R. and Brock, T.D. (1978) 'Life at high temperatures: ecological aspects' in D.J. Kusher (ed.), *Microbial Life in Extreme Environments* (Academic Press, London)

Taylor, D.L., Condeelis, J.S., Moore, P.L. and Allen, R.D. (1973) 'The contractile basis of amoeboid movement. I. The chemical control of motility in isolated cytoplasm.' *Journal of Cell Biology, 59,* 378-94

Taylor, W.D. (1978a) 'Growth responses of ciliate Protozoa to the abundance of their bacterial prey.' *Microbial Ecology, 4,* 207-14

— (1978b) 'Maximum growth rate, size and commonness in a community of bacterivore ciliates.' *Oecologia (Berlin), 36,* 263-72

— (1979) 'Overlap among cohabiting ciliates in their growth responses to various prey bacteria.' *Canadian Journal of Zoology, 57,* 949-51

Taylor, W.D. and Berger, J. (1976) 'Growth responses of cohabiting ciliate protozoa to various prey bacteria.' *Canadian Journal of Zoology, 54,* 1111-14

Thimann, K.V. (1979) 'The development of plant hormone research in the last 60 years' in F. Skoog (ed.), *Plant Growth Substances* (Proceedings of 10th International Conference on Plant Growth Substances, Madison, Wisconsin) (Springer-Verlag, Berlin)

Tibbs, J. (1968) 'Fine structure of *Colpoda steini* during encystment and excystment.' *Journal of Protozoology, 15,* 725-35

Tibbs, J. and Marshall, B.J. (1970) 'Cyst wall protein synthesis and some enzyme changes on starvation and encystment in *Colpoda steini*.' *Journal of Protozoology, 17,* 125-8

Tilney, L.G. and Porter, K.R. (1965) 'Studies on the microtubules of Heliozoa. I. The fine structure of *Actinosphaerium nucleofilium* (Barnett) with particular reference to the axial rod structure.' *Protoplasma, 60,* 317-44

Townsend, C.R. and Hughes, R.N. (1981) 'Maximizing net energy returns from foraging' in C.R. Townsend and P. Calow (eds), *Physiological Ecology* (Blackwell Scientific Publications, Oxford/Sinauer Associates, Inc., Sunderland, Mass.)

Travis, J.L. and Allen, R.D. (1981) 'Studies on the motility of the Foraminifera. I. Ultrastructure of the reticulopodial network of *Allogromia laticollis* (Arnold).' *Journal of Cell Biology, 90,* 211-21

Tucker, J.B. (1968) 'Fine structure and function of the cytopharyngeal basket in the ciliate *Nassula*.' *Journal of Cell Science, 3,* 493-514

— (1974) 'Microtubule arms and cytoplasmic streaming and microtubule bending and stretching of intertubule links in the feeding tentacles of the suctorian ciliate *Tokophrya*.' *Journal of Cell Biology, 62,* 424-37

Uyemura, M. (1936) 'Biological studies of thermal waters in Japan.' *Ecological Review Mount Hokkada Botanical Laboratory, 2,* 171

Vernberg, W.B. and Coull, B.C. (1974) 'Respiration of an interstitial ciliate and benthic energy relationships.' *Oecologia (Berlin), 16*, 259-64

Vickerman, K. (1960) 'Structural changes in mitochondria and *Acanthamoeba* at encystment.' *Nature, London, 188*, 248-9

Walker, G.K., Maugel, T.K. and Goode, D. (1975) 'Some ultrastructural observations on encystment in *Stylonychia mytilus* (Ciliophora: Hypotrichida).' *Transactions of the American Microscopical Society, 94*, 147-54

Wang, C.C. (1928) 'Ecological studies of the seasonal distribution of Protozoa in a freshwater pond.' *Journal of Morphology, 46*, 431-78

Warner, F.D. (1974) 'The fine structure of the ciliary and flagellar axoneme' in M.A. Sleigh (ed.), *Cilia and Flagella* (Academic Press, London and New York)

Warner, F.D. and Satir, P. (1974) 'The structural basis of ciliary bend formation. Radial spoke positional changes accompanying microtubule sliding.' *Journal of Cell Biology, 63*, 35-63

Watson, J.M. (1945) 'Mechanism of bacterial flocculation caused by Protozoa.' *Nature, London, 155*, 271

Watters, C. (1968) 'Studies on the motility of the Heliozoa. I. The locomotion of *Actinosphaerium eichorni* and *Actinophrys*.' *Journal of Cell Science, 3*, 231-44

Webb, M.G. (1956) 'An ecological study of brackish water ciliates.' *Journal of Animal Ecology, 25*, 148-75

—— (1961) 'The effects of thermal stratification on benthic Protozoa in Esthwaite Water.' *Journal of Animal Ecology, 30*, 137-51

Wessenberg, H.S. and Antipa, G. (1970) 'Capture and ingestion of *Paramecium* by *Didinium nasutum*.' *Journal of Protozoology, 17*, 250-70

Wetzel, A. (1928) 'Der fauschlamm und seine ziliaten heitformen.' *Zeitschrift für Morphologie und Okologie der Tiere, 13*, 179-328

Whittaker, R.H. (1969) 'New concepts of the kingdoms of organisms.' *Science, 163*, 150-60

—— (1977) 'Broad classification: the kingdoms and the protozoans' in K. Kreier (ed.), *Parasitic Protozoa* (Academic Press, New York)

Wieser, W. (1973) 'Temperature relations of ectotherms. A speculative review' in W. Wieser (ed.), *Effects of Temperature on Ectothermic Organisms* (Springer-Verlag, Berlin)

Witman, G.B., Plummer, J. and Sander, G. (1978) 'Chlamydomonas flagellar mutants lacking radial spokes and central tubules.' *Journal of Cell Biology, 76*, 729-47

Wohlman, A. and Allen, R.D. (1968) 'Structural organization associated with pseudopod extension and contraction during cell locomotion in *Difflugia.' Journal of Cell Science, 3*, 105-14

Yongue, W.H. and Cairns, J. (1978) 'The role of flagellates in pioneer protozoan colonisation of artificial substrates.' *Polskie Archiwum Hydrobiologii, 25*, 787-801

Zeuthen, E. (1953) 'Oxygen uptake as related to body size in organisms.' *Quarterly Review of Biology, 28*, 1-12

Zinger, J. (1937) 'Zur biologie der infusorien.' *Biologicheskii Zhurnal, 6*, 425-36

INDEX